5th Workshop of Young Researchers in Astronomy and Astrophysics 2009

AA009086

Journal of Physics: Conference Series Volume 218

Budapest, Hungary
2-4 September 2009

ISBN: 978-1-61738-621-3
ISSN: 1742-6588

Printed from e-media with permission by:

Curran Associates, Inc.
57 Morehouse Lane
Red Hook, NY 12571

proceedings.com

Some format issues inherent in the e-media version may also appear in this print version.

Printed by Curran Associates, Inc. (2010)

For permission requests, please contact the Institute of Physics
at the address below.

Institute of Physics
Dirac House, Temple Back
Bristol BS1 6BE UK

Phone: 44 1 17 929 7481
Fax: 44 1 17 920 0979

techtracking@iop.org

Additional copies of this publication are available from:

Curran Associates, Inc.
57 Morehouse Lane
Red Hook, NY 12571 USA
Phone: 845-758-0400
Fax: 845-758-2634
Email: curran@proceedings.com
Web: www.proceedings.com

TABLE OF CONTENTS

REVIEW PAPERS

CONTRIBUTED PAPERS

Preface

The 5th Workshop of Young Researchers in Astronomy and Astrophysics was held on 2–4 September 2009 at the Eötvös University in Budapest, Hungary. This meeting fits into a conference series which can already be considered a tradition where the younger generation has the opportunity to present their work. The event was also a great opportunity for senior astronomers and physicists to form new connections with the next generation of researchers. The selection of invited speakers concentrated on the researchers currently most active in the field, mostly on a post-doctoral/tenure/fresh faculty position level. A number of senior experts and PhD students were also invited.

As the conference focused on people rather than a specific field, various topics from theoretical physics to planetology were covered in three days. The programme was divided into six sections:

- Physics of the Sun and the Solar System
- Gravity and high-energy physics
- Galactic and extragalactic astronomy, cosmology
- Celestial mechanics and exoplanets
- Infrared astronomy and young stars
- Variable stars

We had the pleasure of welcoming 10 invited review talks from senior researchers and 42 contributed talks and a poster from the younger generation. Participants also enjoyed the hospitality of the pub Pál at the Pálvölgyi-cave after giving, hearing and disputing countless talks. Brave souls even descended to the unbuilt, adventurous Mátyásvölgyi-cave.

Memories of the conference were shadowed though. Péter Csizmadia, one of our participants and three other climbers attempted a first ever ascent to the Ren Zhong Feng peak in Sichuan, China, but they never returned from the mountains. Péter departed to China shortly after the conference, with best wishes from participants and friends. We dedicate this volume to his memory.

The organisers thank the Physics Doctoral School of Eötvös University for its hospitality. The workshop was supported by the Mecenatúra and Polányi Mihály Programmes of the National Office for Research and Technology (NKTH) and the Ministry of Education and Culture (OKM) of Hungary.

Emese Forgács-Dajka, Emese Plachy and László Molnár

Obituary for Péter Csizmadia

This conference proceedings is dedicated to the memory of our colleague and friend Péter Csizmadia a young physicist, a computer expert and one of the best Hungarian mountaineers who has been missing since the end of October 2009 and whose last scientific presentation occurred during the 5th Workshop of Young Researchers in Astronomy and Astrophysics, 2–4 September 2009, Budapest.

Péter Csizmadia 1972–2009

The accident: Péter Csizmadia, along with three other young Hungarian mountaineers (Kata Tolnay, Vera Mikolovits, and Balázs Pechtol) disappeared in China's Sichuan region near the Ren Zhong Feng peak of the Himalayas, in the early morning of 23 October 2009. A huge part of a glacier fell, resulting in an extraordinary ice and stone avalanche that covered the valley where they had their camp for the night. Neither their bodies nor their belongings were found by the ground and helicopter rescue teams.

A brief CV: Péter Csizmadia was born in 1972. He received an MSc in physics (1996) and a PhD in heavy ion physics (2003) both of these degrees were issued by Eötvös University (Budapest, Hungary). At the beginning of his career his main research interest was centred around quark hadronization (1995–2005). Later he joined the relativity group of RMKI and worked on numerical simulations in general relativity, investigating problems such as cosmological inflation and gravitational collapse in general relativity (2005–2009).

In 2005, Péter joined to our relativity group and started to work on the first version of a high precision adaptive mesh refinement (AMR) code, called GridRipper. He was the founder and he remained the main developer of GridRipper until his disappearance. Due to his efforts GridRipper became a fourth order precision AMR code implemented as C++/Java classes for solving hyperbolic systems of partial differential equations numerically. The already implemented and tested applications include the study of various dynamical systems such as cosmological inflation, gravitational collapse of spherically symmetric systems, the production and evolution of micro black holes and topology changes. The code is capable of solving hyperbolic partial differential equations (PDE) with any of the usually implemented integration schemes such as second or fourth order Runge–Kutta, Iterated Crank–Nicholson or Lax–Wendroff. The initial data may be given by analytic expressions, by a set of elliptic equations or by numerical data provided by an independent code. The analytic and numerical methods applied in developing GridRipper are unique in the sense that this code – foremost in the World – is capable of following the time evolution inside the black hole region approaching the developing singularity arbitrarily close (see http://www.kfki.hu/~cspeter/numrel/2009-ekg/index.html).

In the meantime András László joined us, and we started to develop GridRipper jointly such that Péter's expertise played a fundamental role in all the new developments. The initial version of GridRipper was suitable for studying only spherically symmetric systems as it was a 1+1 dimensional version. The new version of GridRipper is applicable to generic systems living in 1+N dimensional spacetimes in which the angular degrees of freedom are treated by spectral methods. The effectiveness of the applied method is justified by the fact that simulations that usually require supercomputer background for other methods can be carried out on average PCs by GridRipper.

It was in fact the year 2009 when we reached the desired success, and signs of the true pay-back for the energy invested started to appear. The unexpected disappearance of Péter in China happened when a natural increase of interest within the numerical relativity community became visible. András and I have decided to keep it alive and, if possible, develop GridRipper further, the foundations of which could not be laid without Péter's expertise. This decision is supported not only by the fact that we had the chance to work together with a physicist possessing outstanding programming capabilities, but also because we got to know in Péter a remarkably nice person and a good friend. Peter was always cheerful and ready for discussions or to provide support to anybody who asked for his help. With the loss of Péter we all lost an exceptional colleague and a true friend.

István Rácz

Recent publications of Péter Csizmadia

Csizmadia P and Lévai P 2005 Energy dependence of transverse quark flow in heavy ion collisions *Acta Physica Hungarica* A**22** 371–380

Csizmadia P, 2006 Testing a new mesh refinement code in the evolution of a spherically symmetric Klein–Gordon field *International Journal of Modern Physics* D**15** 107–119

Csizmadia P 2007 Fourth order AMR and nonlinear dynamical systems in compactified space *Class. Quantum Grav.* **24** S369

Hamar G, Zhu L L, Csizmadia P, Lévai P 2008The robustness of quasiparticle coalescence in quark matter *European Physical Journal Special Topics* **155** 67–74

Hamar G, Zhu L L, Csizmadia P, Lévai P 2008 Strange hadron yields and ratios in heavy ion collisions at RHIC energy *J. Phys. G: Nucl. Part. Phys.* **35** 044067

Csizmadia P and Rácz I 2008 On the cosmological relevance of oscillons, Poster at the *Frontiers in Numerical Gravitational Astrophysics Summer School*, Erice, Italy, June 27–July 5

Csizmadia P, Rácz I 2010 Gravitational collapse and topology change in spherically symmetric dynamical systems *Class. Quantum Grav.* **27** 015001

All papers published in this volume of *Journal of Physics: Conference Series* have been peer reviewed through processes administered by the proceedings Editors. Reviews were conducted by expert referees to the professional and scientific standards expected of a proceedings journal published by IOP Publishing.

Journal of Physics: Conference Series **218** (2010) 012001 doi:10.1088/1742-6596/218/1/012001

Measuring and modelling light pollution at the Zselic Starry Sky Park

Zoltán Kolláth
Konkoly Observatory, Konkoly Thege u. 15-17, H-1121 Budapest, Hungary
E-mail: kollath@konkoly.hu

Abstract. One of the first 'International Dark-sky Parks' in Europe was established at the Zselic Landscape Protection Area in Hungary. A special monitoring program has been carrying on to survey the quality of the night sky using 'Sky Quality Meters' and DSLR cameras. The main conclusion of our measurements is that the local villages have only a minimal effect on the quality of the sky. There are light-domes due to the neighbouring cities only close to the horizon, the main source of obtrusive light is the city of Kaposvár. The anthropogenic component of zenith luminance of the night sky is obtained as the function of the distance from the city centre of Kaposvár. Our data were modelled by radiation transfer calculations. These results can help to draw attention to the energy emitted useless to the space and to protect our nocturnal landscape of nature parks for the next generations.

1. Introduction

At the Zselic Landscape Protection Area, Hungary an extensive program has been started to conserve the excellent night sky quality of the region. The International Dark-Sky Association announced the designation of two additional Dark Sky Parks on November 15, 2009: Galloway Forest Park in Scotland, UK and the Zselic Landscape Protection Area in Hungary, simultaneously became the first European International Dark Sky Park. Recognition as an International Dark Sky Park requires intense long term planning, retrofitting and preservation efforts to fulfil the application prerequisites.

The Zselic region, which is located in the South-West part of Hungary, is one of the best locations for dark skies in the county. The area of the protected park is more than 9000 hectares, and its major part is woodland. The Zselic Landscape Protection Area has excellent night sky quality, on clear nights the artificial component of the night sky brightness does not exceed the natural luminance of the sky. The Milky Way, the zodiacal light and other faint phenomena are clearly visible by naked-eye on a clear night.

Artificial lighting, when poorly designed, does not lit only the desired area and a large amount of light out of this area is usually wasted. This has energetic consequences hence the wasted luminous flux has to be paid by the community, and the production of the wasted energy has again negative effects on our environment. On the other hand our environment cannot only be polluted by emission of different materials but also by undesired light. The wasted luminous flux makes its effect somewhere else, e.g. it can cause serious changes in the natural environment and ecosystem and increases the luminance of the night sky. To determine this effect on a nature park, night sky luminance measurements and radiation transfer calculations should be performed.

The US National Park Service started first monitoring of the night sky at natural protected areas like national parks in 2001 (Duriscoe, Luginbuhl & Moore 2007). These data are also used for the protection of the nocturnal visual resource. This system also allows comprehensive characterization of sky conditions at astronomical observatories. Luginbuhl et al. (2009) used the same type of observations to test the predictions of light pollution models. In this paper we show our methods of measurements used in the Zselic Starry Sky Park and the interpretation of our data.

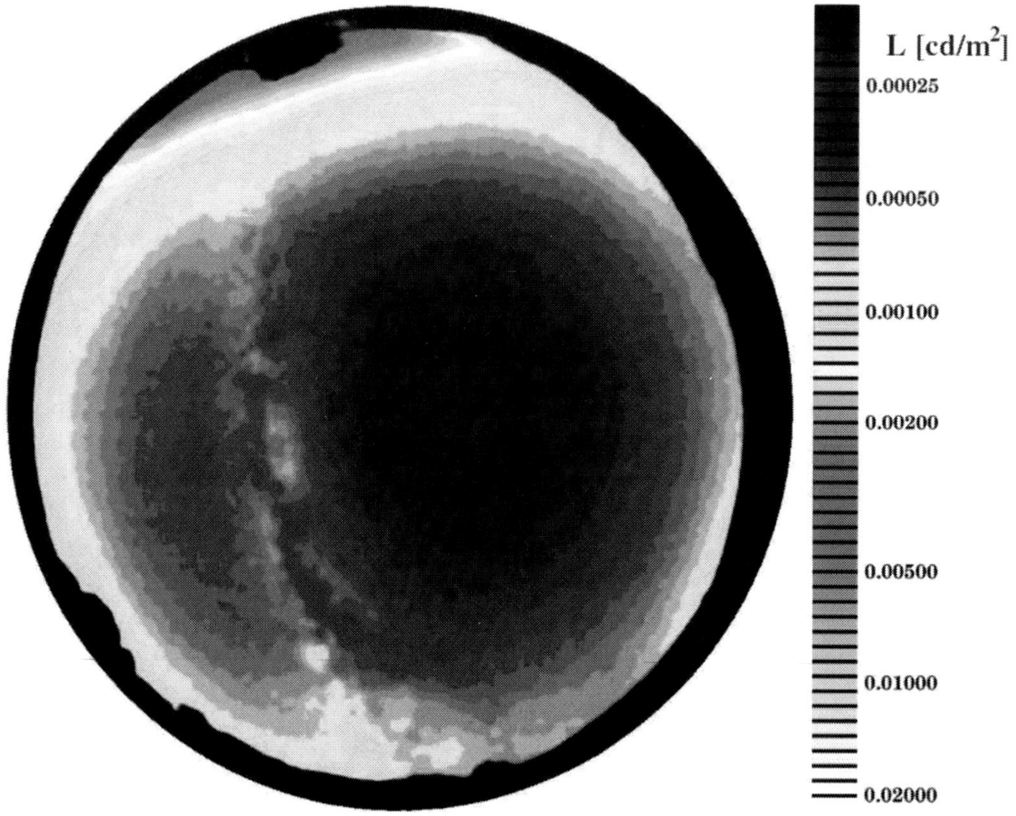

Figure 1. Luminance distribution of the clear night sky at a typical location inside the Zselic Landscape Protection Area. The brighter part at the top of the image indicates the direction of the city of Kaposvár.

2. Observational methods

There are plenty of methods to estimate the quality of the night sky. Most of them are visual estimates of the limiting magnitude (the faintest star which can be visible by naked eyes) or the number of stars observable in a fixed region of the sky. These methods work well to describe the expected view of the night sky; however, they depend on subjective factors, like the eye-sight of the observer. To map and monitor the light pollution of the park and its neighbourhood it was necessary to find methods to measure the quality of the sky in a reproducible way with calibrated instruments. The natural quantity that describes the 'darkness' of the nocturnal environment is the luminance of the sky.

It is expected that the luminance of the sky varies on a distance scale of a few kilometres, so it was necessary to perform a field survey to map the distribution if sky brightness thorough the park area and in its neighbourhood. The geographic co-ordinates of the measurements were registered with a precision of at least 100 meters. At normal conditions there are no detectable changes of zenith luminance on this distance scale. We used a GPS device or a detailed topographic map of the region to get the geographic co-ordinates. The locations were recorded in Universal Transverse Mercator (UTM) co-ordinates. The UTM co-ordinates can be practical when one uses topographic map with this scale, and it also provides the distances between measurement locations and the exact directions of polluting light sources easily. When needed, UTM data are converted to standard longitude-latitude data.

2.1. Luminance measurements

We selected a portable photometer with high enough sensitivity to measure the subtle light of the rural night sky. A simple device, the Sky Quality Meter (SQM) by Unihedron was designed for this purpose, and it is used widely for light pollution monitoring. SQM is a handheld device; it collects the light from a relatively large solid angle (1.5 steradian, approximately a cone with a 42 degrees half angle). The device displays the average luminance of this solid angle in astronomical units: magnitude per square arcsecond (mag/arcsec^2). SQM is temperature calibrated and gives the luminance with the precision of 0.1 mag/arcsec^2, which is equivalent to 10 percent in linear luminance (cd/m^2) units. Laboratory tests of this instrument were presented by Cinzano (2007).

We have been monitoring the sky quality using SQM since January 27, 2007. All of our measurements were taken with the device pointing to the zenith. The necessary conditions to perform a field survey were the followings:

- Moonless night.
- No clouds or fog.
- The Sun is at least 18 degrees below the horizon (astronomical twilight).
- No direct light from artificial sources reaches the detector of the device.

2.2. Measurements by all sky imaging

Recent Digital Single Lens Reflex (DSLR) cameras provide a new opportunity to monitor the quality of the night sky and light pollution. Cameras that are able to save images in an unaltered raw format, can be calibrated to get measurements of the luminance of the sky in a physical scale. Then the photo of the night sky can be converted to false colour images, which represents the distribution of sky luminance. Our experiments show that commercial DSLR cameras can easily reach 10 percent precision (e.g. 0.1 mag/arcsec^2), when calibrated thoroughly.

Our standard settings for light pollution monitoring in dark locations are the followings:

- Canon EOS 350D camera (some parallel tests with EOS 50D, 450D and 300D)
- Sigma 4.5mm f2.8 EX DC circular fisheye lens
- ISO 800 setting
- T=180 s exposure time
- f= 2.8-3.5 aperture

We developed our own software (dclum) to process raw images and to calibrate the camera. The camera and the lens were calibrated in co-operation with the University of Pannonia at Veszprém and the Technical College of Budapest. The system was first calibrated in a photometry lab and then with a high performance imaging luminance meter under real sky conditions. To get as precise calibration as possible, we used different types of standard measurements and then compared the results: The laboratory measurements were used also to fit the effect of vignetting of the lens. The calibration was performed in standard photopic luminance scale (unit of cd/m^2). Then the measurements can be transformed to mag/arcsec^2 units with a good approximation. An example of the resulting luminance map is displayed on Figure 1.

We have cross correlated the DSLR measurements integrated with the sensitivity curve of SQM with parallel SQM measurements. The results are satisfactory as they agree within 0.1 mag/arcsec^2 (10%). We also checked the linearity of the CCD by multiple exposures and different light levels. It is concluded that the raw images of a digital camera can be used for our purposes since the calibration errors do not exceed the precision needed for light pollution monitoring.

The precision of our treatment is clearly demonstrated with a simple experiment: The light-dome of Budapest was photographed from a vantage point at a distance of 80 km from the Capital (Piszkéstető Mountain Station, Konkoly Observatory). A 5% decrease of the city's light dome is clearly measurable during the switch off of the ornamental lights in Budapest.

3. Models

To interpret the sky brightness measurements and to provide models of the light pollution in the park we have developed a Monte-Carlo radiation transfer code. In clear air (no clouds), the propagation of light is determined by Rayleigh scattering for molecules and by Mie scattering for aerosols. Absorption is negligible in the visible range of electromagnetic radiation. The mean free path of photons in terrestrial atmosphere is several tens of kilometres, depending on the aerosol content and elevation. It gives a natural choice to use Monte-Carlo simulation of the light propagation the observed photometric quantities are statistical averages of photon packets. For detailed description of Monte-Carlo radiation transfer calculations see e.g. Spada et al. 2006.

To take into consideration the effect of different humidity and aerosol content we calculated model sequences with different zenith extinction. A standard atmospheric model for clear air was used, and we fitted the aerosol content to get a fixed zenith extinction. In this way we obtained a one parameter models sequence which well represents the behaviour of the atmosphere in clear (cloudless) nights.

4. Measurements

The main conclusion of our measurements is that the local settlements have only a minimal effect on the quality of the sky. The luminance is only slightly increased in the vicinity of small villages. The lights from the city of Kaposvár, North of the area mostly affect the sky background. Inside the Landscape Protection Area the quality of the night sky is better than 21.0 mag/arcsec^2, typically around 21.3-21.4 mag/arcsec^2. There is a clear increase in night sky luminance North of the park area, but acceptable conditions can be found for stargazing even at those locations. To display the quantitative results, we plotted the luminance of the night sky, as the function of the distance from the city centre of Kaposvár (Figure 2). For this plot we converted the SQM measurements to the standard luminance scale (cd/m^2). It provides a more natural way to subtract the background sky brightness (L_0), which represents the natural sky luminance and the overall sky brightness from distant artificial sources. SQM measurements are plotted for a night with snow coverage at the region of Kaposvár (red dots), and for selected nights with similar atmospheric quality (blue crosses).

We also performed Monte Carlo atmospheric radiation transfer calculations to interpret our measurements. On Figure 2. we displayed the trends from these calculations with solid lines. The upper line represents twice as large luminance values as the lower curve does. Both curves are calculated for clear (low humidity and aerosol content) air.

With L_0=0.00032 cd/m^2 the distribution of measurements can be well represented with the models. This value agrees with our darkest measurements (21.5-21.6 mag/arcsec^2) in the region. There is a clear overall trend in sky luminance measurements, agreeing with the radiation transfer

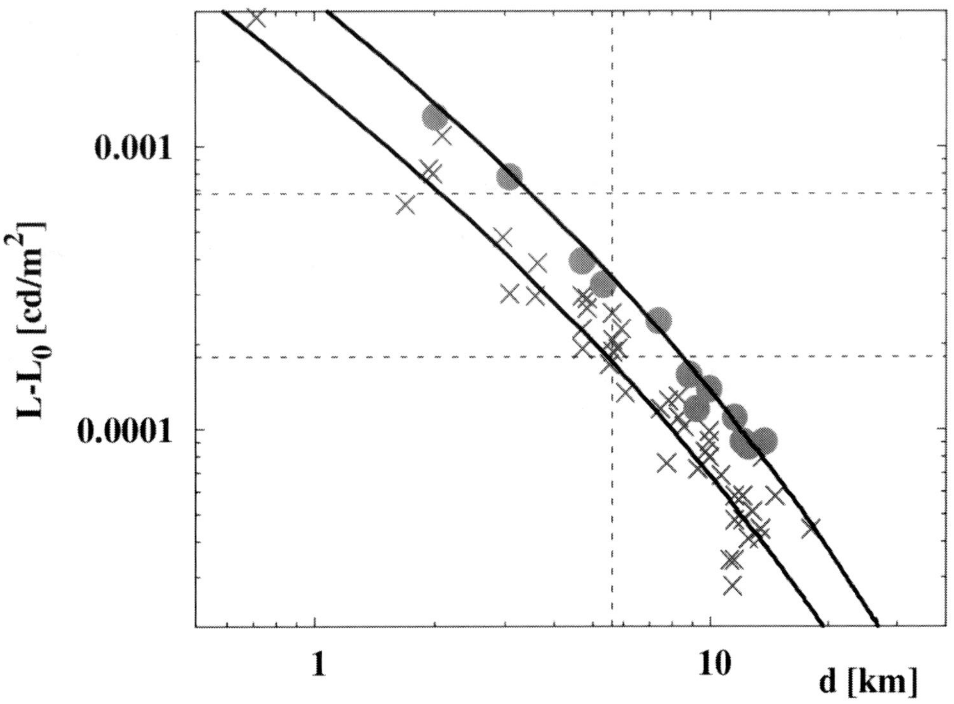

Figure 2. Artificial sky brightness as a function of the distance from the city of Kaposvár. Blue dots: normal ground conditions; red dots: snow coverage; lines: radiation transfer models.

calculations. The deviations from this trend are due to the measurements being taken close to local polluting sources, to changes in weather conditions and to instrumental noise. The dotted horizontal lines indicate the L=0.00025 and 0.00075 cd/m^2 (equal to, and three times larger than the natural sky luminance). The vertical line shows the distance of the closest border of the LPA to the city. We can conclude that the quality of the sky inside the Zselic Landscape Protection Area is above the limits of silver tier of the International Dark-Sky Association, even at snowy conditions.

There is another interesting conclusion one can draw from the above measurements. The increased reflexivity of the ground by snow coverage doubled the luminance of the sky. It suggests that a significant component of light pollution inside the LPA originates from ground reflection. Taking into consideration that the ground reflection increased from the normal 0.1-0.2 values to the numbers related to the mixture of snow and wet road (0.6-0.8), the ratio of direct to reflected components of sky brightness can be estimated. These calculations show, that up to 50% of the sky glow due to Kaposvár could be eliminated, if only fully shielded fixtures were used in Kaposvár. The measurements together with the radiation transfer calculations can be used, to argue for a Lighting Master Plan in Kaposvár.

The luminance calibrated images are displayed on a false colour scale to display the luminance distribution on the sky. The major obtrusive sources can be clearly identified when such images (after a mirroring) are displayed on a map of the region. It is also possible to get the spatial dependence of sky quality inside the Landscape protection area.

Based on all sky luminance distributions, it is easy to determine different single quantities, like a simulated SQM value (SQM_{sim}), or the total sky brightness above 20 degrees (TSB_{20}, see e.g. Duriscoe, Luginbuhl & Moore 2007). These quantities also demonstrate the excellent night time quality of the Zselic region.

The spatial distribution of the luminance of the sky is visualized with better clarity, if the circular map is transformed to the plane of azimuth and elevation. These images show that the light dome of Kaposvár is limited to elevations below 20 degrees even from a closest location (Figure 3).

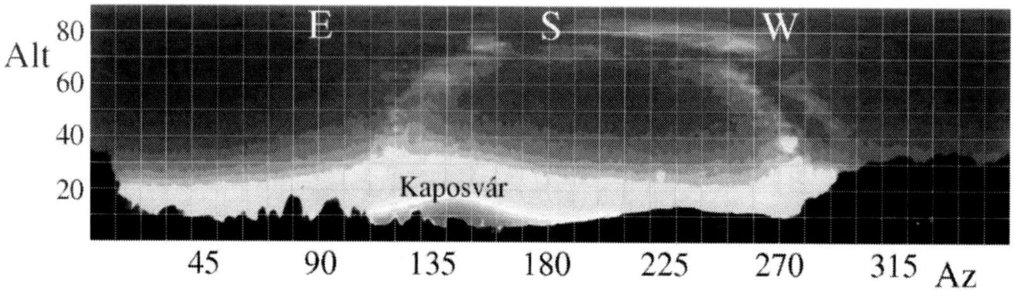

Figure 3. Luminance distribution of the sky as a function of azimuth and altitude. The same colour scale is used as in Figure 1.

5. Conclusion

We have performed sky quality (night sky luminance) measurements in the Zselic Landscape Protection Area and its vicinity. It is demonstrated that both the Sky Quality Meter and a DSLR camera are suitable for taking quick luminance measurements in remote locations. The pocket size, handheld device is especially useful for mapping light pollution at multiple locations during a one-night long monitoring session. All sky cameras (e.g. DSLR cameras with circular fisheye lens) give additional information in sky quality. The main light pollution source in the park is the city of Kaposvár. Our sky background measurements show that the light pollution due to these light sources is reduced to acceptable level in the distance of 7 km from the city centre and these lights do not have any serious effect at the Zselic Starry Sky Park region. Local settlements have only very limited effect on night sky quality. Everywhere inside the landscape protection area the quality of the sky is better than the minimal requirements for the silver tier recommended by the International Dark Sky Association.

We fitted the sky luminance measurements as a function of distance from the centre of Kaposvár by the results of radiation transfer calculations of light pollution. With the models it is also possible also predict any degradation or improvement in sky quality related to changes in the artificial light sources in the neighbourhood. We plan to use these predictive models to negotiate with municipalities, stakeholders and designers of lighting systems in the region in order to avoid any increase of light pollution at the Dark Sky Preserve. We continue monitoring the sky quality to notice any changes that need corrective steps. We plan to establish a grid of fixed measuring stations in the Zselic region to provide automatic and continuous monitoring of the luminance of the night sky.

References

Cinzano P 2007 *ISTIL Internal Report* **9**
Duriscoe D M, Luginbuhl C B, Moore C A 2007 *PASP* **119** 192
Luginbuhl C B, Duriscoe D M, Moore C W, Richman A, Lockwood G W, Davis D R 2009 *PASP* **121** 204
Spada F, Krol M C, Stammes P 2006 *Atmospheric Chemistry and Physics Discussions* **6** 4823

Journal of Physics: Conference Series **218** (2010) 012002 doi:10.1088/1742-6596/218/1/012002

The dynamical solar atmosphere

István Ballai[1] **and Emese Forgács-Dajka**[2]

[1]Solar Physics and Space Plasmas Research Centre, Dept. of Applied Mathematics, University of Sheffield, Hounsfield Road, Hicks Building, Sheffield, S3 7RH, UK

[2]Eötvös University, Department of Astronomy, H-1518 Budapest, Pf. 32., Hungary

E-mail: i.ballai@sheffield.ac.uk; e.forgacs-dajka@astro.elte.hu

Abstract. High resolution space and ground-based telescopes made it clear that one of the most fundamental characteristics of solar atmosphere is that it is dynamic, showing stationary and non-stationary motion on all time and space scales. Dynamical changes in the solar atmosphere are rather difficult to interpret given the presence of the magnetic field and the high-degree of ionization of the plasma. After all, the magnetic field is the quantity which controls the dynamics, stability, and thermal state of the plasma. The present contribution aims to review some fundamental knowledge on plasma dynamics in the solar atmosphere with special emphasis on coronal seismology, i.e. the science which is using the observable dynamical changes in the solar corona to infer quantities that cannot be measured (structure and magnitude of the coronal magnetic field, scale-heights, density filling factors, etc.).

1. Introduction

High resolution observations made in the last decade have changed dramatically our perception about the dynamical phenomena occurring in different layers of the solar atmosphere. Waves, which until mid 90s were sporadically observed (mainly in radio frequency), have become now a reality, being observed in all sort of wavelengths. The new achievement also meant that new areas of solar physics can have now an observational background.

The dynamical response of the plasma in the solar atmosphere to rapid changes in the solar interior can be manifested through waves propagating within the solar atmosphere. The dynamics of a large class of waves with wavelengths and periods large compared with the ion Larmor radius (10^{-4} m in the solar photosphere and about 1 m for almost all combinations of coronal parameters) and the gyroperiod (10^{-7} s in the solar photosphere and of the order of 10^{-3} s in the solar corona), respectively, can be described within the framework of collisional magnetohydrodynamics (MHD). Waves perturb macro-parameters of the plasma, such as density, temperature, bulk velocity and magnetic field. The period range of waves and oscillations observed in the solar atmosphere (from a few seconds to a few hundreds of minutes) is well covered by temporal resolution of presently available ground-based and spaceborne observational tools. However, it is very important to understand that in normal circumstances both spatial and temporal resolution are necessary ingredients of successful detection of the waves, e.g. the pixel size must be much smaller than the wavelength and the temporal cadence time must be shorter than the period. Only in certain exceptional cases, e.g. when the wave passes through a bright object with the geometrical size smaller than the wavelength, the wave can be detected with poorer spatial resolution.

The importance of studying waves and oscillations in the solar atmosphere resides on at least three aspects. First of all MHD waves are able to transport energy and momentum from the solar deeper regions up to the chromosphere and corona, therefore contributing to the thermalization of the upper solar atmosphere and/or acceleration of the solar wind. Secondly waves and oscillations are able to transport information about the medium in which they propagate, therefore seismological techniques are used to obtain information about the solar interior, solar atmosphere and the coupling between different regions in the solar atmosphere. Finally, waves and oscillations are important as they are able to trigger some other secondary effects (very important though) in space plasmas, such as magnetic reconnection, instabilities, turbulences, etc.

Given the abundance of results, a full review of waves and oscillations including observation, theory and modelling could be a very challenging and time consuming task, different aspects were summarized previously in great detail (see, e.g. Banerjee et al 2007 for observations and Nakariakov and Verwichte 2005 for theory and modelling).

2. MHD waves in solar plasma – an observational overview

Ordinary sound waves in air owe their existence to the restoring force created by a gas pressure gradient. If we introduce a magnetic field to this system we create two more restoring forces, the magnetic tension and magnetic pressure gradient. If we assume that the plasma is homogeneous (every element of the plasma is identical to any other element) and infinite, the ideal MHD equations predict the existence of three different types of waves called the slow magnetoacoustic, Alfvén and fast magnetoacoustic waves (if gravitational and non-inertial forces are neglected).

The slow magnetoacoustic (or just simply slow) waves are driven by the magnetic and plasma pressure interacting destructively, whereas the fast magnetoacoustic (FMA) waves are produced when the magnetic and gas pressure interact constructively. The Alfvén wave has an intermediate wave speed, and is entirely driven by the magnetic tension. A polar diagram of the phase speeds of the three types of waves is given in Fig 1. It is obvious that slow waves can only travel at very small angles off parallel to the magnetic field lines and fast waves travel in any direction, with a larger speed when they propagate perpendicular to the ambient magnetic field lines. Alfvén waves have a preference to transport energy along field lines and never across them (analogous to vibrations on a guitar string).

The presence of waves and oscillations in the solar atmosphere is strongly connected to the magnetic field; one cannot fully understand the possible spectrum of waves without a detailed knowledge of the magnitude and structure of the magnetic field. Waves are generated, in general, by buoyouncy forces (gravity, magnetic field, pressure gradients, Coriolis forces, etc.) but also due to the convection motion in the sub-surface region and transient and energetic phenomena occurring at different heights in the solar atmosphere. Waves and oscillations can be categorized in many ways, however for our purposes we can distinguish between local and global waves depending whether the wave is connected or not to a certain magnetic structure. Although they may seem separate phenomena, in fact they are very much related in the sense that often global waves are able to generate local waves and oscillations.

The solar atmosphere is a highly structured medium, with large gradients created by the magnetic field. The magnetic field is not dispersed uniformly, instead it tends to accumulate in entities known as flux tubes and this imposes a strict ordering of solar magnetism. The process of emergence of flux tubes is believed to be caused by the massive convective motions below the photosphere.

In the photosphere one of the most obvious magnetic feature is the sunspot typified by a cool dark region, with strong (3kG) magnetic field, often with a 10^4 km diameter with temperatures of about 4000 K, about 2000 K cooler that sunspots' surrounding. Sunspots are not homogeneous, but generally exhibit an irregular pattern of bright points, called umbral dots, some 150 km in

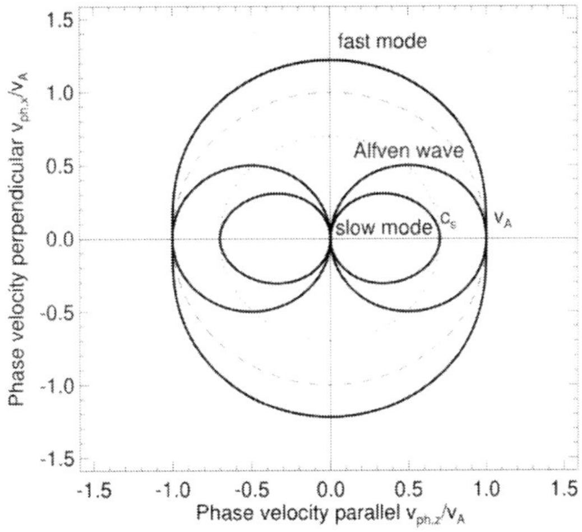

Figure 1. Polar plot of the phase speeds of the three types of MHD waves. The magnetic field lines lie parallel to the horizontal axis. Here the sound speed (c_S) is 70% of the Alfvén speed (v_A). The Alfvén speed and sound speed are given by the dashed and dotted line, respectively

diameter with brightness similar to that of the surrounding 6000 K photosphere and with an average lifetime of 1500 s. Magnetic field in these dots are expected to be somehow less then in their surroundings.

It is well-know that several oscillations are connected to the sunspots, most important being the 180 s (chromosopheric oscillation) and the 300 s (photospheric oscillation). Observations of sunspots oscillations have been summarised (see, e.g. Lites et al. 1982; Brynildsen et al. 2000) as comprising vertical oscillations within the umbrae (with periods of 130-190 s), and horizontal penumbral waves running outwards at speeds of 10-20 km s^{-1} and periods of 210-300 s. Apart from sunspots, the rest of the photosphere is far from being a homogeneous medium. It is now generally accepted that the magnetic field over most of the solar surface is concentrated into regions of magnetic flux ranging from small ($\sim$ 100 km, 1 kG) flux tubes, through knots (500 km, 1.5 kG) and pores ($\sim 10^3$ km, 2 kG) to sunspots.

It now widely agreed that the magnetic flux concentrations of the photosphere (usually at the supergranules boundaries) appear to be co-spatial with magnetic features higher in the solar atmosphere, the chromosphere. Briefly, the principal features of the chromosphere may be classified as horizontal ($\leq$ 300 km above the limb) fibrils which appear to outline flux tubes joining areas of opposite polarity and spicules which are radial jets of gas which shoot upwards with average velocities of 20 km s^{-1} from the chromospheric network along magnetic field lines. Recently several attempts have been made to explain spicules as the effect of p-mode buffeting on photospheric network of magnetic field lines or reconnection of opposite polarity magnetic field lines (see, e.g. De Pontieu et al. 2004, Pasachoff et al. 2009) while the multitude of waves and oscillations in spicules has been reviewed recently by Zaqarashvili and Erdélyi (2009).

X-ray and (E)UV observations by high resolutions space satellites (SOHO, TRACE, HINODE) have changed our view of the dynamical state of the corona. Since the Skylab mission we know that the corona consists of multitude of high emissivity coronal loops with different lengths, temperature, density, etc. at very high temperature (majority have temperature in excess of 1 MK). Since the plasma is controlled by magnetic forces (in corona the plasma-beta is much smaller than 1), coronal loops are tracers of coronal magnetic field lines. Loops may vary in their characteristics over a wide range. The small X-ray bright points which have time scales of the order of 8 hours are thought to be loop structures while coronal loops believed to form large prominences, some extending 100 Mm above the limb, have an average lifetime of a few days to a few months. Some of the loops are "closed" being anchored by magnetically opposite footpoints in the denser chromosphere and photosphere, extending as high as 300 Mm whereas others are "open" fading high in the corona and coinciding with the source of the high-speed solar wind. Values for magnetic field inside coronal loops were obtained using extrapolation techniques from the photosphere but also with seismological methods revealing that intensities vary between a few G up to 100 G resulting in Alfvén speeds of the order of 1000-2000 km s^{-1}. Emission line widths also revealed that coronal loops are denser structures than their environment, the filling factor (density ratio between inside and outside the loop) varying between 5 and 10. These values usually refer to loops in active regions, being the easier to observe, especially those appearing shortly after a solar flare.

Given the complex structure of coronal loops it is expected that these magnetic entities will support a rich variety of waves and oscillations. One of the most studied type of waves is the so-called kink wave which propagates along a magnetic flux tube so that the symmetry axis of the tube is distorted (see, e.g. Aschwanden et al. 1999, Nakariakov et al. 1999, Douglas et al. 2010). It is believed that kink waves and oscillations are the result of the interaction between an external driver (possibly a global wave or nearby sudden energy release) and the coronal loop (Selwa et al. 2006, Ogrodowczyk and Murawski 2007, Hindman and Jain 2008, Ballai et al. 2009). Kink oscillations in coronal loops and their very rapid damping allowed the estimations of magnetic fields, density scale-heights, sub-resolution structure, heating functions, etc. (see, e.g. Nakariakov et al. 1999, Andries et al. 2005; Verth et al. 2008, Douglas et al. 2009). On the other hand, if the oscillations occur such that the symmetry axis of loops is not dislocated we are dealing with sausage modes and they were observed already in coronal loops by, e.g. Aschwanden et al. (2003), Erdélyi and Taroyan (2008). Both of these waves are belonging to the family of magnetoacoustic waves which arise due to the combined effect of forces due to the magnetic tension and pressure gradients.

Finally we should mention that apart from magnetoacoustic waves, magnetic structures can support pure magnetic waves (Alfvén waves) which always propagate along magnetic field lines, are transversal and very little influenced by non-ideal effects. Since Alfvén waves are incompressible they will not show any variation of intensity in EUV wavelengths. The only way to observe Alfvén waves is by Doppler line broadening. Although the propagation speed and characteristics of Alfvén waves are very much similar to the kink waves, the observation of Alfvén waves is very sporadic (e.g. Harrison et al. 2002, Jess et al. 2009), however future observational facilities will provide the required resolution for observing Alfvén waves.

Apart from waves propagating in various magnetic structures, the solar atmosphere permits the propagation of waves which are able to cover a very large extent of the solar surface, i.e. global waves. One of the most known global waves are the so-called p-modes propagating in the solar photosphere and can be considered as acoustic modes able to penetrate the solar interior. These oscillations were first discovered by Leighton (1960). p-modes are generated by the convection motion inside the Sun where convective cells are continuously buffeting the convectively stable photosphere (Stix 1970). It is believed that the solar photosphere is covered by thousands of p-modes at all times. The power spectrum of these standing oscillations shows a maximum peak at about 300 s. Higher frequency p-modes are observed in the chromosphere. Their velocity amplitude lies between 0.1 and 1.6 km s^{-1} in the photosphere but it is increasing with height. The horizontal wavelength of these oscillations is observed to be between 500 to 10,000 km, the higher limit being larger than a typical size of a granule. Vertical and horizontal speeds lie in the interval 30 to 100 km s^{-1}. The great advantage of p-modes is that they can carry information about the structure (density, temperature, chemical composition, etc.) and dynamical state (shear meridional and poloidal flows, stability, etc.) of the solar interior. Helioseismology is the science which studies the properties of p-modes for solar interior diagnostics.

The other family of global waves consists of large-extent waves that are generated by sudden and powerful energy releases (flares/CMEs). Impulsively driven global waves have been known since the early 1960s. Although it is still not known how the release of energy and energized particles will transform into waves, today it is widely accepted that these disturbances are similar to the circularly expanding bubble-like shocks after atomic bomb explosion or shock waves which follow the explosion of a supernova. Thanks to the available observational facilities, global waves were observed in a range of wavelengths in different layers of the solar atmosphere. A pressure pulse can generate seismic waves in the solar photosphere propagating with speeds of 200 – 300 km s^{-1} (Kosovichev and Zharkova 1998; Donea et al. 2006). Higher up, a flare generates very fast super-alfvénic shock waves known as *Moreton waves* (Moreton and Ramsey, 1960), best seen in the wings of H_α images, propagating with speeds of 1000 – 2000 kms^{-1}. In the corona, a flare or CME can generate an EIT wave (Thompson et al. 1999) first seen by the SOHO/EIT instrument or an X-ray wave seen in SXT (Narukage et al. 2002). There is still a vigourous debate how this variety of global waves are connected (if they are, at all). Co-spatial and co-temporal investigations of various global waves have been carried out but without a final widely accepted result being reached.

Unambiguous evidence for large-scale coronal impulses initiated during the early stage of a flare and/or CME has been provided by the Extreme-ultraviolet Imaging Telescope (EIT) observations onboard SOHO and by TRACE/EUV. EIT waves propagate in the quiet Sun with speeds of 250 – 400 km s^{-1} at an almost constant altitude. At a later stage in their propagation EIT waves can be considered a freely propagating wavefront which is observed to interact with coronal loops (see, e.g. Wills-Davey and Thompson 1999). Using TRACE/EUV 195 Å observations, Ballai et al. (2005) have shown that EIT waves (seen in this wavelength) are waves with average periods of the order of 400 seconds. Since at this height, the magnetic field can be considered vertical, EIT waves were interpreted as fast MHD waves. This conclusion is further supported by other observations (see, e.g. Long et al. 2008, Patsourakos et al. 2009)

3. MHD waves in solar plasma – a theoretical overview

All waves in which we are interested in the present review can be described within the framework of MHD, which is a set of highly nonlinear equations combining the Navier-Stokes equation with Maxwell system of equations. The general solution of the MHD system of equation is complicated but various realistic assumptions may be made to assist the analysis. First of all, we assume that all waves in which we are interested are linear, so perturbations of physical quantities are much smaller than their equilibrium value. Secondly, since here we are interested in the basic characteristic of waves, we will assume that the plasma is ideal, so free from any non-ideal effects. Further simplifications can be made by considering the location of waves' appearance. For example, in the magnetically dominated corona the so-called *zero-beta* (or cold plasma) approximation involving the neglect of sound speed compared with the Alfvén speed, is an acceptable simplification. In addition when wavelengths are smaller than the density scale-height, gravitational effects (i.e. inhomogeneity along the direction of propagation) can be also neglected. This latter assumption is valid for those coronal waves which are observed to propagate for a rather limited length, but not possible for oscillations involving the whole coronal loop. Since the density scale-height decreases from the corona to photosphere in the lower regions of the solar atmosphere gravitational effects will be more important. One property common to the variety of solar phenomena mentioned above which can be often exploited mathematically is that the transversal dimensions of magnetic structures are much smaller than longitudinal ones and as a result the so-called *slender flux tube* approximation can be used.

Almost all waves and oscillations presented earlier are strictly connected to the magnetic structure in which they propagate, therefore the magnetic flux tube acts as a waveguide (that is why these waves can be labelled as local waves). The fact that these solar features have lateral boundaries means that the usual waves of an infinite medium – presented earlier – are modified by the structuring.

3.1. MHD waves in structured media

In order to investigate the effect of structuring in plasmas we assume that the structure has infinite extent in the z-direction of a Cartesian coordinate system and the effect of gravitational stratification is neglected. In the equilibrium state we suppose that the plasma pressure, p, density, ρ, and magnetic field induction $B\hat{\mathbf{z}}$ are all x dependent. In equilibrium the total pressure, made up of gas pressure, $p(x)$, and magnetic pressure, $B^2(x)/2\mu$, must be constant, i.e.

$$\frac{d}{dx}\left(p+\frac{B^2}{2\mu}\right)=0, \tag{1}$$

where μ is the permeability of free space.

Linear and isentropic perturbations for velocity (with components $\mathbf{v}=(v_x,v_y,v_z)$) and magnetic field ($\mathbf{b}=(b_x,b_y,b_z)$) about the equilibrium state (1) may then be Fourier-decomposed in space and time by writing the variables in the form

$$v_x=\hat{v}_x(x)e^{i(\omega t+ly+kz)}, etc., \tag{2}$$

where the x-dependent amplitude of the velocity component, $\hat{v}_x$ satisfies

$$\frac{d}{dx}\left[\frac{\rho(x)(k^2v_A^2(x)-\omega^2)}{m^2(x)+l^2}\frac{\hat{v}_x}{dx}\right]-\rho(x)\left[k^2v_A^2(x)-\omega^2\right]\hat{v}_x=0, \tag{3}$$

where $m^2(x)$ is the magnetoacoustic parameter defined as

$$m^2(x)=\frac{(k^2c_S^2(x)-\omega^2)(k^2v_A^2(x)-\omega^2)}{(c_S^2(x)+v_A^2(x))(k^2c_T^2(x)-\omega^2)}. \tag{4}$$

In the above equations the sound, c_S, Alfvén, v_A, and cusp (tube), c_T, speeds are defined as

$$c_S(x)=\left[\frac{\gamma p(x)}{\rho(x)}\right]^{1/2}, \quad v_A(x)=\frac{B(x)}{[\mu\rho(x)]^{1/2}}, \quad c_T(x)=\frac{c_S(x)v_A(x)}{[c_S^2(x)+v_A^2(x)]^{1/2}}, \tag{5}$$

with γ being the adiabatic index. Eq. (3) is the Cartesian equivalent of the Hain-Lüst equation arising in cylindrical geometry and has been obtained earlier by Goedbloed (1971), Chen and Hasegawa (1974), Wentzel (1979), Roberts (1981).

A general solution of Eq. (3) is a rather difficult task due to the singularities of the equation occurring when $\omega^2=k^2v_A^2$ (Alfvén singularity) and $\omega^2=k^2c_T^2$ (slow singularity). Due to the transversal inhomogeneity, the nature of modes changes dramatically. While homogenous plasmas have a spectrum of linear eigenmodes which can be divided into slow, fast and Alfvén subspectra, with the slow and fast subspectra having discrete eigenmodes and the subspectrum of Alfvén waves being infinitely degenerated, in inhomogeneous plasmas the three subspectra are changed. The infinite degeneracy of the Alfvén point spectrum is lifted and replaced by the Alfvén continuum along with the possibility of discrete Alfvén modes occurring, the accumulation point of the slow magnetoacoustic eigenvalues is spread out into the slow continuum and a number of discrete slow modes may occur, and the fast magnetoacoustic point spectrum accumulates at infinity. Eigenfunctions that correspond to frequencies in the continuum spectra are improper as they contain a non-square integrable singularity at the two singular positions. The continuous spectra appearing in these plasma allows the resonant interaction between plasmas and external waves if the frequency of the external wave matches any frequency in the slow and/or Alfvén continuum.

The resonant coupling of system is always accompanied by a bi-directional exchange of energy; so that if the energy of the external wave is transferred to the plasma, we speak about resonant absorption while if the energy flow is directed from the plasma to the wave, we are dealing with instabilities. The process of resonant interaction of waves and the possibility of energy transfer between global and local waves has been studied intensively for the past few decades. First resonant absorption was studied as a means of supplementary heating fusion plasmas, but was later rejected due to technical difficulties (see, e.g. Tataronis and Grossman 1973, Chen and Hasegawa 1974, Poedts et al. 1989, van Eester et al. 1991).

In the Earth's magnetosphere resonant MHD wave coupling is believed to generate low frequency pulsation or energize ULF waves (see, e.g. Southwood and Hughes 1983, Taroyan and Erdélyi 2003). Within the context of solar physics Ionson (1978) was the first who proposed resonant absorption as a possible mechanism for coronal heating. His idea was further developed in the last few decades by many authors (see, e.g. Kuperus et al. 1981, Hollweg 1991, Goossens 1991, Sakurai et al. 1991, Goossens and Ruderman 1995, Ofman and Davila 1995, Ballai et al. 1998ab, Ballai and Erdélyi 1998, Ballai et al. 2000, Ruderman and Roberts 2002, etc.), however it became clear soon that resonant absorption alone cannot explain the very high temperature of the corona. Resonant absorption was also used to explain the loss of power in p-modes when interacting with sunspots in the solar photosphere (see, e.g. Goossens and Poedts 1992, Spruit and Bogdan 1992, Tirry et al. 1998).

Recently resonant absorption acquired a new application in *coronal seismology* where the rapid damping (with damping times of the order of a few periods) of kink oscillations in coronal loops is explained in terms of resonant interaction of global kink modes and the local Alfvén waves (Ruderman and Roberts 2002, Goossens et al. 2006, Terradas et al. 2010). Coronal seismology is dependent on theoretical relations (dispersion relations) which link plasma parameters, such as the plasma density, to wave parameters, such as the wave frequency, in a precise way. Generally, plasma parameters are determined from wave parameters, which themselves are determined observationally. The dispersion relations for many complicated plasma structures under the assumptions of ideal MHD are well known; they were derived long before accurate EUV observations were available using simplified models within the framework of ideal and linear MHD. Although the realistic interpretation of many observations are made difficult by the spatial and temporal resolution of present satellites not being adequate, considerable amount of information about the state of the plasma and the structure and magnitude of the coronal magnetic field have been already obtained.

When the medium is discretely structured, comprising slabs and cylinders of piecewise uniform magnetic field and density, Eq. (3) can be solved in a straightforward way. Let us consider first a slab of magnetic field, having a width x_0, with equilibrium described by quantities with an index "0" for the plasma inside the slab ($|x| < x_0$) and by quantities having an index "e" describing the plasma outside the slab ($|x| > x_0$). The structure is taken as infinite in the y-direction, and for simplicity we restrict our analysis to two-dimensional analysis, i.e. v_y=0 and l=0. To model waves and oscillations in magnetic structures, let us consider waves that are confined to the inhomogeneity $|x| < x_0$, thus the inhomogeneity acts as a wave guide with disturbances outside the slab ($|x| > x_0$) being laterally evanescent, i.e. $\hat{v}_x \to 0$ as $|x| \to \infty$. In this case the solutions of Eq. (3) are given by

$$\hat{v}_x(x) = \begin{cases} \alpha_e e^{-m_e(x-x_0)}, & x > x_0, \\ \alpha_0 \cosh m_0 x + \beta_0 \sinh m_0 x, & |x| < x_0, \\ \alpha_e e^{m_e(x+x_0)}, & x < -x_0, \end{cases} \tag{6}$$

where α_0, β_0, α_e and β_e are arbitrary constants, while m_0 and m_e are values of the magnetoacoustic parameter, m, in the two regions. In order to satisfy the condition of evanescence of waves outside the slab we also require that $m_e > 0$. Solutions of Eq. (3) are obtained by applying conservations laws at the boundaries of the slab, i.e. at $|x| = x_0$ we impose the continuity of the normal component of the velocity and total pressure (magnetic plus kinetic).

For an incompressible plasma, corresponding to $\gamma \to \infty$, both m_0 and m_e tend to $|k|$ so the explicit solution of Eq. (3) can be written simply as

$$\frac{\omega}{k} = \pm \left[\frac{\rho_0 v_A^2 + \rho_e v_{Ae}^2 \tanh |k| x_0}{\rho_0 + \rho_e \tanh |k| x_0} \right]^{1/2}, \tag{7}$$

for symmetric modes (also called sausage modes) and

$$\frac{\omega}{k} = \pm \left[\frac{\rho_0 v_A^2 + \rho_e v_{Ae}^2 \coth |k| x_0}{\rho_0 + \rho_e \coth |k| x_0} \right]^{1/2}, \tag{8}$$

for asymmetric or kink modes. The solutions of Eqs. (7) and (8) are Alfvénic surface waves which owe their existence to the presence of boundaries and their amplitude attain the maximum value at the boundaries and decline on either side of $x = \pm x_0$.

In the case of waves with their wavelength much longer than the width of the slab (i.e. $kx_0 \ll 1$) Eqs. (7) and (8) reduce to

$$\omega = \pm k v_A \left\{ 1 + \frac{\rho_e}{\rho_0} \left(\frac{v_{Ae}^2}{v_A^2} - 1 \right) |k| x_0 \right\}^{1/2}, \tag{9}$$

for sausage waves and to

$$\omega = \pm k v_{Ae} \left\{ 1 + \frac{\rho_e}{\rho_0} \left(\frac{v_A^2}{v_{Ae}^2} - 1 \right) |k| x_0 \right\}^{1/2}, \tag{10}$$

for the kink wave. In a wide slab, i.e. $|k| x_0 \gg 1$, both modes have phase speed given approximately by $\omega^2 = k^2 c_K^2$, where

$$c_K^2 = \frac{\rho_e v_{Ae}^2 + \rho_0 v_A^2}{\rho_e + \rho_0}, \tag{11}$$

is known as the kink speed and it can be regarded as a density weighted Alfvén speed.

Returning to the compressible case, the four constants in Eq. (6) may be eliminated by applying the continuity conditions at the two interfaces, obtaining

$$\rho_e (k^2 v_{Ae}^2 - \omega^2) m_0 \left\{ \begin{array}{c} \tanh \\ \coth \end{array} \right\} m_0 x_0 + \rho_0 (k^2 v_A^2 - \omega^2) m_e = 0, \tag{12}$$

where the two cases correspond to sausage and kink modes, respectively. Since Eq. (12) is a transcendental equation, it possesses a rich spectrum of solutions and within the context of solar plasma physics it has been first derived by Roberts (1981). Although a full solution of this equation might look complicated, specific applications to, e.g. photospheric or coronal structures allow an easier interpretation. According to the standard nomenclature, solutions corresponding to $m_0^2 > 0$ are labelled as surfaces waves, while waves with $m_0^2 < 0$ are body waves. This difference in solutions refers to the spatial behaviour inside the structure; surface waves have their maximum amplitude on the boundary are are not oscillatory inside the slab, while body waves show an oscillatory pattern inside the slab.

In the unstructured plasma one expects three groups of waves, the slow and fast sets of magnetoacoustic waves and the set of transversal Alfvén waves. Our analysis has uncoupled the latter, so these pure magnetic waves will not be consider further but we must note that the slow and fast waves met in infinite plasmas are altered in a complicated way depending on the relative magnitudes of sound and Alfvén speed inside and outside the slab. In some cases, for example, that of a cold plasma, in which $v_{Ae} > v_A$ (often applied for coronal structures) only two sets of body waves occur whereas in photospheric applications where $v_{Ae} < v_A$, the slow mode can always propagate (either as a surface or body mode) but the fast mode may only propagate in a structure that is cooler than its surrounding (i.e. $c_e > c_0$).

4. Waves in coronal structures

In the solar corona the dynamics is controlled by the magnetic field, here the plasma-beta parameter is much less than 1, therefore it is right to use the so-called low-beta approximation ($\beta = 2c_S^2/\gamma v_A^2 \ll 1$) or cold-plasma approximation ($\beta = 0$). In the cold plasma approximation only fast waves will survive. Writing $n_0^2 = -m_0^2$, Eq. (12) for fast sausage body and fast kink body mode reduces to

$$\tan(n_0 x_0) = -\frac{n_0}{m_e}, \qquad \tan(n_0 x_0) = \frac{m_e}{n_0}. \tag{13}$$

In the cold plasma limit, n_0 and m_e are given by

$$n_0 = \frac{\omega^2 - k^2 v_A}{v_A^2}, \qquad m_e^2 = \frac{k^2 v_{Ae}^2 - \omega^2}{v_{Ae}^2}. \tag{14}$$

It is interesting to note that the equations given by Eq. (13) have arisen in a number of other, physically dissimilar, situations. The first equation of Eq. (13) arises in seismological and oceanographical studies (see, e.g. Pekeris 1948) while the second equation has been earlier obtained within the context of elastic layers of the Eath's crust (also known as Love waves, Love 1911).

It is also interesting to consider the case of a cylindrical inhomogeneity. For simplicity we will assume that $c_e = c_0 = 0$ and the plasma is described within the low-beta plasma would give qualitatively similar results. For a cylinder of radius a and plasma density ρ_0 surrounded by a plasma of density ρ_e, all embedded in a uniform field $B\hat{\mathbf{z}}$, waves of the form

$$v_r = \hat{v}_r(r) e^{i(\omega t + n\theta + kz)}, \tag{15}$$

satisfy the dispersion relation

$$\frac{J_n'(n_0 a)}{J_n(n_0 a)} \frac{K_n(m_e a)}{K_n'(m_e a)} = -\frac{n_0}{m_e} \tag{16}$$

and J_n and K_n are Bessel functions of order n with their derivatives J_n' and K_n' (Edwin and Roberts 1983).

The behaviour of phase speed ω/k as a function of the dimensionless wavenumber ka is shown in Figure 2, allowing for the effects of non-zero sound speeds (but keeping the low-β approximation). These modes are guided waves of a region of low Alfvén speed (i.e. $v_A < v_{Ae}$), corresponding to a region of high density ($\rho_0 > \rho_e$). If the Alfvén speeds inside the loop would be higher than outside, that would give rise to radial propagation in the outside cylinder, i.e. waves become leaky. In coronal loops there are no surface waves, i.e no solutions are found for $n_0^2 < 0$, however, as Figure 2 shows, there are two sets of body modes widely separated.

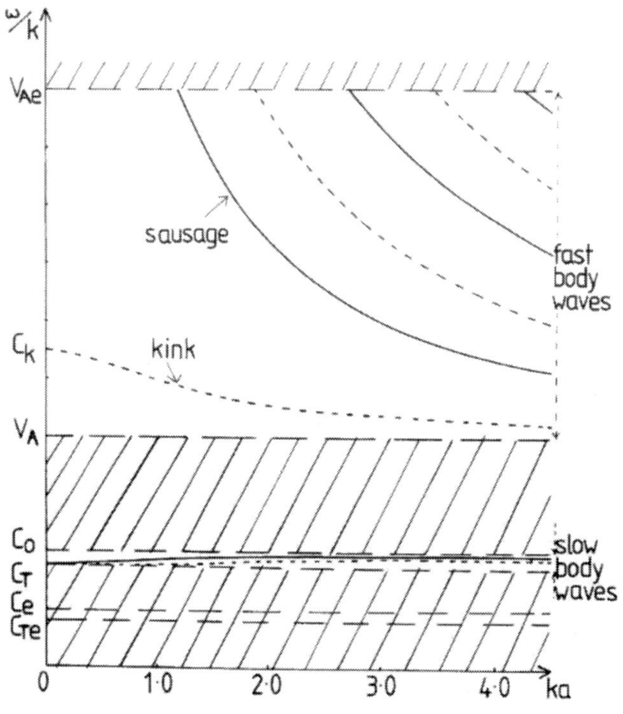

Figure 2. The phase speed ω/k for magnetoacoustic modes in terms of the dimensionless wavenumber ka. The above dispersion curves are obtained for typical coronal conditions ($v_{Ae} > v_A > c_S > c_T$). Sausage modes and kink modes are indicated by solid and dashed lines, respectively (adopted from Edwin and Roberts 1983).

The lower band corresponds to slow waves (nearly sound waves) constrained to propagate one dimensionally along the almost rigid field lines. These waves are weakly dispersive in a slender tube ($ka \ll 1$) and they propagate with the frequency

$$\omega \approx kc_T \approx kc_0 \qquad \text{for} \quad c_0 \ll v_A. \tag{17}$$

The upper band of Fig. 2 corresponds to fast oscillations of high density and are highly dispersive except the fundamental kink mode which has a phase speed close to c_K.

In a closed coronal loop with its footpoints anchored in the dense photosphere it is likely that standing modes occur. One of the mostly studied coronal oscillations are the kink oscillations mentioned earlier generated by the interaction of coronal loops with external drivers. The nature of possible standing modes in coronal loops was recently studied by Ballai et al. (2009) where the analysis was carried out for a wide range of drivers. The results show that the signal recovered in the loop is always a superposition of loops' natural oscillations and the characteristics of the driver. For a loop of length L and $k = j\pi/L$ the times scales involved in the analysis for slow/fast waves are

$$\tau_S \approx \frac{2L}{c_0}, \qquad \tau_f \approx \frac{2L}{c_K} \tag{18}$$

For typical coronal sound speed of 200 km s^{-1} and and kink speeds of 1000 km s^{-1} for a loop length of 100 Mm we would obtain periods of about 1000 sec for slow waves and 200 sec for kink modes, such scales being already observed.

5. Waves in photospheric structures

Since temperatures in the solar photosphere are much lower than in the corona, gravitational effects are more important here, appearing as stratification in equilibrium quantities, i.e. all equilibrium quantities can depend on the vertical coordinate along the gravitational force. As specified earlier, the structuring manifests in the form of flux tubes from readily visible sunspots to the sub-resolution intense tubes. The lower solar atmosphere is divided into strong-field media (the intense flux tubes) or field-free media (the tubes' surrounding) though within sunspots themselves there is fine structure in the form of umbral dots. Therefore some progress in describing possible waves in these structures can be made by using Eq. (12) with $v_{Ae} = 0$ or $v_A = 0$. In order to simplify the description but keep the essential physics we are going to suppose that we are dealing with slender tubes where we restrict our analysis to waves of long wavelength compared to the width of the structure. The effect of density and pressure decrease with height is that as we move upward the cross-section area of the tube is increasing however the flux within the tube is always conserved.

The equations describing the dynamics of longitudinal and isentropic motion (characterised by $v(z,t)$) of a gas of density $\rho(z,t)$ and pressure $p(z,t)$ confined within an elastic tube of cross-sectional area $A(z,t)$ can be given as

$$\frac{\partial \rho A}{\partial t} + \frac{\partial}{\partial z}(\rho v A) = 0, \tag{19}$$

$$\rho\left(\frac{\partial v}{\partial v} + v\frac{\partial v}{\partial z}\right) + \frac{\partial p}{\partial z} + \rho g = 0, \tag{20}$$

$$\frac{\partial p}{\partial t} + v\frac{\partial p}{\partial z} = \frac{\gamma p}{\rho}\left(\frac{\partial \rho}{\partial t} + v\frac{\partial \rho}{\partial z}\right), \tag{21}$$

$$p + \frac{B^2}{2\mu} = p_e, \qquad BA = \text{const.} \tag{22}$$

The last two equations express that the tube will be stable if the total pressure inside the tube is balanced by the external pressure and that the magnetic flux is conserved at all times. Here the gravity is assumed to be aligned with the z-axis of the tube ($\mathbf{g} = -g\hat{\mathbf{z}}$) and p_e is the gas pressure in the tube's exterior.

The simplest way to deal with the above system of equation is to linearize the equations assuming small amplitude waves so we will assume that all quantities (except speed) can be written as the sum of an equilibrium value and its small perturbation. The equilibrium quantities can be written as

$$p_0(z) = p_0(0)e^{-n_0(z)}, \quad \rho_0(z) = \rho_0(z)\frac{\Lambda_0(0)}{\Lambda_0(z)}e^{-n_0(z)},$$
$$A_0(z) = A_0(0)e^{n_0(z)/2}, \qquad B_0(z) = B_0(0)e^{n_0(z)/2}, \tag{23}$$

where

$$n_0(z) = \int_0^z \frac{dz'}{\Lambda_0(z')}, \qquad \Lambda_0(z) = \frac{p_0(z)}{\rho_0(z)g},$$

is the pressure scale-height (here we assumed that the temperatures inside and outside the tube are identical).

Assuming that p_e remains equal to its equilibrium value, it may be shown that the re-scaled longitudinal component of the velocity

$$v(z,t) = \frac{Q(z,t)}{[\rho_0(z)A_0(z)c_T^2(z)]^{1/2}}, \tag{24}$$

is governed by the Klein-Gordon equation (Webb and Roberts 1978, Rae and Roberts 1982, Ballai et al. 2006)

$$\frac{\partial^2 Q}{\partial t^2} - c_T^2(z)\frac{\partial^2 Q}{\partial z^2} + \omega_v^2 Q = 0. \tag{25}$$

The general expression of ω_v is rather cumbersome but in the case of an isothermal atmosphere (Λ_0=const.) it reduces to

$$\omega_v = \frac{c_T}{\Lambda_0}\left[\frac{9}{16} - \frac{1}{2\gamma} + \left(\frac{\gamma-1}{\gamma}\right)\frac{c_0^2}{v_A^2}\right]^{1/2}. \tag{26}$$

Thus the geometry and elasticity of a magnetic flux tube give rise to a cut-off frequency, ω_v, which is analogous to the cut-off frequency, $\omega_a = c_0/2\Lambda_0$, for vertically propagating acoustic-gravity waves in a field-free medium. The presence of the cut-off frequency implies that only frequencies above the cut-off propagate under adiabatic conditions; frequencies below this value are evanescent. In the light of this statement, it is obvious that only those magnetoacoustic waves can propagate towards the chromosphere and corona whose frequency is above the cut-off value. The solution of the Klein-Gordon equation shows that under the effect of stratification a slow wave will propagate along field lines with speed c_T followed by a wake oscillating with the frequency ω_v.

In the chromosphere, due to the decreases density, many waves will have their amplitude increased and they can easily evolve into nonlinear waves (shocks or solitons), a subject which is not going to be discussed here. It is interesting to note that the Klein-Gordon equation presented earlier seems to have an universal character, the same equation was found to describe the propagation of externally driven kink oscillations in coronal loops (Ballai et al. 2009).

Strictly speaking the corona and solar wind are not fully collisional media, therefore the MHD system of equation may not be the best formalism to describe waves. The derivation of the isotropic MHD equations requires the assumption that electrons are tightly coupled to ions, so that plasma temperatures are isotropic. In this case the plasma density and temperature satisfy some equation of state. However, spacecraft observations often show anisotropic velocity and temperature distributions, implying that this model is not fully valid for many space plasmas. In this kind of plasma, different types of particle can have different temperatures parallel and perpendicular to the ambient magnetic field. Thus, more complex models based on more sophisticated closure relations may be necessary to model adequately the rich diversity of plasma dynamics.

Parker (1958) was the first to point out that the interplanetary magnetic field might lead to an anisotropic pressure tensor. When the cyclotron frequency of the ions is much higher than the collisional frequency, the particles gyrate many times around a line of magnetic force between two collisions. The magnetic field induces a splitting in the pressure, introducing a parallel and a perpendicular component. The two components of the pressure are not necessarily equal. However, the thermal anisotropy and electron conduction velocity relative to the ions generally excite plasma oscillations that scatter the particles, pulling the plasma toward isotropy, ($p_\parallel \sim p_\perp$).

The main complication arising from the presence of anisotropy relates to pitch angle scattering, a mostly kinetic effect. Such a description within fluid approximation remains a fundamental but unresolved theoretical problem. Technically speaking, the closure of the fluid equations requires two equations of state, one for the parallel and one for the perpendicular pressure. In spite of much effort, there is no generally accepted form for these two equations of state. In such a medium waves that are mostly generated by the free energy of the particle distribution functions are the means to relax this free energy. Waves, once amplified, can heat particles, permit exchange of energy between different populations of particles and precipitate particles.

An original approach to the problem of physical properties of rarefied plasmas has been the theory developed by Chew, Goldberger and Low (1956) known as the Chew-Goldberger-Low (CGL) approximation. The usual MHD equations are derived from the Boltzmann equation using an expansion in powers of the collisional mean free path. In this case the plasma is collision-dominated therefore the collisional term in the Boltzmann equation is the leading term, with all other terms being treated as perturbations. When the density is so small that the plasma can be considered as a (near) collisionless medium, a new form of governing equations can be derived from the Boltzmann equation using an expansion in powers of the Larmor radius. This approximation can be considered as an adiabatic approximation since it depends on the Larmor frequency being large compared to other characteristic frequencies of the problem.

The properties of anisotropic plasmas have been studied intensively over the past few years in the light of new observations (see, e.g. Kato, Tajiri and Taniuti 1966; Hau and Lin 1995; Nakariakov and Oraevsky 1995; Ballai et al. 2002, Ballai and Zhugzhda 2002, Ballai and Marcu 2004). These studies revealed that the properties of the waves are different from the usual MHD and these differences are crucial from the wave propagation and stability point of view. Since the study of waves in anisotropic plasmas is in its infancy there is no general concept as to how these waves are generated. However, it is very likely that the generation mechanism of these waves in similar to the mechanisms that generate the well-known MHD (isotropic) waves: convection, reconnection, resonances with other modes, even non-anisotropic, etc. For example, a wave that propagates in the isotropic plasma can 'decay' into a two-component wave (i.e. parallel and perpendicular sound speeds).

Thanks to the high resolution observations our knowledge about the dynamical state of the solar atmosphere is ever changing, however every observation will help us understand how our star is actually working.

Acknowledgments

I. Ballai acknowledges the financial support by NFS Hungary (OTKA under grant no. K67746) and The National University Research Council Romania (CNCSIS-PN-II/531/2007) and the great effort of the local organizers to make the meeting a time to remember. E. Forgács-Dajka acknowledges the financial support by the Hungarian Science Research Fund (OTKA) under grant no. K67746 and no. K81421 and by the EC Solaire Network (MRTN-CT-2006-035484).

References

Andries J, Arregui I, and Goossens M 2005 *Astrophys. J* **624** L57
Aschwanden M J, Fletcher L, Schrijver C J and Alexander D 1999 *Astrophys. J.* **520** 880
Aschwanden M J 2003 in *NATO Advanced Research Workshop* eds. Erdélyi et al., Budapest, Hungary
Ballai I, Ruderman M S and Erdélyi R 1998a *Phys. Plasmas* **5** 252
Ballai I, Erdélyi R and Ruderman M S 1998b *Phys. Plasmas* **5** 2264
Ballai I and Erdélyi R 1998c *Solar Phys.***180** 65
Ballai I, Erdélyi R and Goossens M 2000 *J. Plasma Phys.* **64** 235
Ballai I, Erdélyi R, Voitenko Y and Goossens M 2002 *Phys. Plasmas* **9** 2593
Ballai I and Zhugzhda Y D 2002 *Phys. Plasmas* **9** 4280
Ballai I and Marcu A 2004 *Astron. Astrophys.* **415** 691
Ballai I, Erdélyi R and Pintér B 2005 *Astrophys. J.* **633** L145
Ballai I, Erdélyi R and Hargreaves J 2006 *Phys. Plasmas* **13** 042108
Ballai I, Douglas M and Marcu A 2009 *Astron. Astrophys.* **488** 1125
Banerjee D, Erdélyi R, Oliver R, O'Shea E 2007 *Sol. Phys.* **246** 3
Brynildsen N, Maltby P, Leifsen T, Kjeldseth-Moe O and Wilhelm K 2000 *Sol. Phys.* **191** 129
Chen L and Hasegawa A J 1974 *J. Geophys. Res.* **79** 1033
Chew G F, Goldberger M L and Low F E 1956 *Proc. Roy. Soc. London A* **236** 112
De Pontieu B, Erdélyi R, James S P 2004 *Nature* **430** 536

Donea A-C, Besliu-Ionescu D, Cally P, Lindsey C and Zharkova V V 2006 *Solar Phys.* **239** 113
Douglas M, Ballai I and Jess D B 2009 *Astron. Astrophys.* in press
Edwin P and Roberts B 1983 *Sol. Phys.* **88** 179
Erdélyi R and Taroyan Y 2008 *Astron. Astrophys.* **489** 49
Goedbloed J P 1971 *Physica* **53** 412
Goossens M 1991 *Advances in Solar System Magnetohydrodynamics* ed Priest E R and Hood A W (Cambridge: Cambridge University Press)
Goossens M and Poedts S 1992 *Astrophys. J.* **384** 348
Goossens M and Ruderman M S 1995 *Phys. Scripta* **T60** 171
Goossens M, Andries J and Arregui I 2006 *Phil. Trans. Roy. Soc. A.* **364** 433
Harison R A, Hood A W and Pike C D 2002 *Astron. Astrophys.* **392** 319
Hau L-N and Lin C A 1995 *Phys. Plasmas* **2** 294
Hindman B W and Jain R 2008 *Astrophys. J.* **677** 899
Hollweg J V 1991 *Mechanisms of Chromospheric and Coronal Heating* ed Ulmschneider P, Priest E R and Rosner R (Berlin: Springer)
Ionson J A 1978 *Astrophys. J.* **226** 650
Jess D B, Mathioudakis M, Erdélyi R, Crockett P J, Keenan F P and Christian D J 2009 *Science* **323** 1582
Kato Y, Tajiri M and Taniuti T 1966 *J. Phys. Soc. Jpn* **66** 765
Kosovichev A G and Zharkova V 1998 *Nature* **393** 317
Kuperus M, Ionson J A and Spicer D 1981 *ARA&A* **19** 7
Leighton R B 1960 *IAU Symp.* **12** 321
Lites B W, White O R and Packman D 1982 *Astrophys. J.* **253** 386
Long D M, Gallagher P T, McAteer R T J and Bloomfield D S 2008 *Astrophys. J.* **680** L81
Moreton G F and Ramsey H E 1960 *Publ. Astron. Soc. Pac.* **72** 357
Nakariakov V M, Ofman L, DeLuca E E, Roberts B and Davila J M 1999 *Science* **285** 862
Nakariakov V M, Verwichte E 2005 *Living Rev. Sol. Phys.* **2** 3
Nakariakov V M and Oraevsky V N 1995 *Sol. Phys.* **158** 29
Narukage N, Hudson H S, Morimoto T, Akiyama S, Kitai R et al. 2002 *Astrophys. J.* **572** L109
Ogrodowczyk R and Murawski K 2007 *Astron. Astrophys.* **461** 1133
Parker E N 1958 *Phys. Rev.* **109** 1874
Pasachoff J M, Jacobson W A and Sterling A C. 2009 *Sol. Phys.* **260** 59
Poedts S, Kerner W and Goossens M 1989 *J. Plasma Phys.* **42** 27
Patsourakos S, Vourlidas A, Wang Y M, Stenborg G and Thernisien A 2009 *Sol. Phys.* **259** 49
Rae I C and Roberts B 1982 *Astrophys. J.* **256** 761
Roberts B 1981 *Sol. Phys* **69** 27
Ruderman M S and Roberts B 2002 *Astrophys. J.* **577** 475
Sakurai T, Goossens M and Hollweg J V 1991a *Solar Phys.***133** 227
Selwa M, Solanki S K, Murawski K, Wang T J and Shumlak U 2006 *Astron. Astrophys.* **454** 653
Southwood D J and Hughes W J 1983 *Space Sci. Rev.* **35** 301
Spruit H C and Bogdan T J 1992 *Astrophys. J.* **391** 109
Stix M 1970 *Astron. Astrophys.* **4** 189
Taroyan Y and Erdélyi R 2003 *J. Geophys. Res.* **108** 1043
Tataronis M and Grossman M 1973 *Zeitschrift für Physik* **261** 223
Terradas J, Goossens M and Ballai I 2010 *Astron. Astrophys.* in press
Tirry W J, Goossens M, Pintér B, Cadez V and Vanlommel P 1998 *Astrophys. J.* **503** 422
Thompson B J, Gurman J B, Neupert W M et al. 1999 *Astrophys. J.* **517** 151
Verth G, Erdélyi R, and Jess D B 2008 *Astrophys. J* **687** L45
Zaqarashvili T V and Erdélyi R 2009 *Space Sci. Rev.* in press
van Eester D, Goossens M and Poedts S 1991 *J. Plasma Phys.* **45** 3
Webb A R and Roberts B 1978 *Sol. Phys.* **59** 249
Wentzel D G 1979 *Astrophys. J.* **227** 319
Wills-Davey M J and Thompson B J 1999 *Solar Phys.* **190** 467

5th Workshop of Young Researchers in Astronomy & Astrophysics IOP Publishing
Journal of Physics: Conference Series **218** (2010) 012003 doi:10.1088/1742-6596/218/1/012003

Notes on exoplanets

Szilárd Csizmadia

Deutsches Zentrum für Luft- und Raumfahrt, Institut für Planetenforschung, 12489 Berlin, Rutherfordstrasse 2., Germany

E-mail: szilard.csizmadia@dlr.de

Abstract. Our knowledge about exoplanets evolves rapidly. Here I give a short overview of some aspects of the exoplanet research and I also introduce shortly the reader to the Hungarian activities in the exoplanet field.

1. Introduction

This note summarizes the main points of the talk I had given in the 5th Workshop of Young Researchers in Astronomy and Astrophysics, held at Eötvös University, Budapest, Hungary. The talk was an introductionary talk to the session 'Celestial Mechanics and Exoplanets' in which I reviewed the methods of planet detection, the discovery statistics, transit spectroscopy results, and I listed the observed secondary eclipses and the most important properties of exoplanets. Because of lack of space, the present note does not reflect the whole content of that talk.

One can found recent, more detailed overviews about the properties of exoplanets in Udry & Santos (2007), Santos (2008), Baraffe, Chabrier & Barman (2010), Schneider (2010) and Swift et al. (2010).

2. Notes on exoplanets

2.1. Number of known exoplanets and exoplanetary systems

The following statistics is based on Schneider (2010). Up to now (2010 Feb 28) we know 430 confirmed planets, 69 unconfirmed or controversial planets and six free-floating planets. Most of the exoplanets (339 out of 430) were discovered by the so-called radial velocity method. Although this method gives only the lower mass limit for the companion object, we can assume that the orbits of these planets are really close to the line of the sight and they remain in the planetary mass range even if the inclination is little bit lower than 90 degrees. Nine planets were discovered by the timing method which also has the same problem, i.e. gives only the lower mass limit for the suspected planetary mass object. With other methods (including microlensing, direct imaging, etc.) further nine objects were detected. 70 objects show transits: their orbits are oriented in space such a way that a certain small part of the orbit is in front of the stellar disc, therefore we can observe periodically that the planet moves in front of the apparent stellar disc covering a certain part of the star and causing a decrease in the observable stellar flux. The detailed analysis of the flux variation – combining with spectroscopic measurements – can tell us the radius as well as the orbital inclination of the planet and the planet-star distance, too. In this case we are able to determine the real mass of the planet and hence its mean density which is important for studies of the internal structure and evolution of planets. As one can easily see, the chance to have a transit is quite small (it is proportional to $1/a$, where a is the semimajor axis of the planet orbit: the more distance the planet from the star, the less the chance to observe it by transit method). That is why it is not surprising, that 67 transiting exoplanets have orbital period shorter than 10 days and there are only three known exoplanets among the 70 known such ones which have orbital period longer than 10 days (HD 17156b: $P = 21.22$ days, Fisher et al. (2007), HD 80606b: $P = 111$ days Moutou et al. (2009) and CoRoT-9b: $P = 95$ days, Deeg et al. (2010) – only this latter one was discovered by transit method, the two other ones were detected by radial velocity method first and the transits were observed subsequently).

2.2. The smallest known exoplanets

There is a natural high interest in Earth-like exoplanets. The currently known smallest exoplanets belong to the class of 'Super-Earths' (planets with 1-10 Earth-masses, see Valencia, O'Connell & Sasselov 2006). So far the smallest known exoplanet is CoRoT-7b, which was detected by its transits from space using the CoRoT satellite, and it has 1.7 Earth-radius and 4.8 ± 0.8 Earth-masses (Léger et al. 2009, Queloz et al. 2009). Interestingly, the average density of this exoplanet is quite similar to the one of the Earth (Queloz et al. 2009), suggesting that its internal structure can be similar to the Earth's one. It seems that CoRoT-7b, which orbits its K-type host star during only 20 hours (the orbital radius is only ~ 4 times larger than the stellar radius) cannot be a remnant of a hot-Jupiter or hot-Neptune which would lost its envelope due to the radiation of the host star (Lammer et al. 2009). An interesting question arises in Barnes et al. (2010): is CoRoT-7b a real Super-Earth or is it a super-Io? If it were a super-Io, then – heated tidally by its host star – it would show volcanism (Barnes et al. 2010). Other authors state that CoRoT-7b can be a remnant of a former ocean-planet (Wagner et al. 2009).

The surface gravity on this exoplanet is 1.4-2 times larger than on the surface of the Earth (depending on its mass). The surface temperature in the substellar point is about 1800-2600 K and it decreases down to ~ 50 K in the coolest point on the night side (Léger et al. 2009). However, the temperature profile from the hottest point to the coolest one is thought to be continuous, so somewhere the temperature is between 0-50 °C. Theoretically, there can be water in this region, but it seems to be unlikely. Of course, this is true if and only if the planet rotational period is synchronized with the orbital period. However, one can assume that this is the case, because tidal locking will occur on the timescale (Guillot et al. 1996):

$$t_{lock} = \frac{2\pi Q_{star}}{P_{rot}} \frac{R^3_{star}}{GM_{star}} \frac{M^2_{star}}{M_{planet}} \left(\frac{a}{R_{star}}\right)^6 \tag{1}$$

where $Q_{star} \sim 10^5$ (Guillot et al. 1996), P_{rot} is the rotational period of the star, G is the gravitational constant, M_{star} and R_{star} are the stellar mass and radius, a is the planet's orbital semimajor axis, and M_{planet} is the planet's mass. Substituting the values of CoRoT-7b system taken from Léger et al. (2009) and (Queloz et al. 2009) we have that the tidally locked spinning of the planet occurs within 50-60 Myrs. Since the stellar age is between 1.2-2.3 Gyrs (Léger et al. 2009), we can assume that the planet is already tidally locked (however, it is difficult to investigate this issue observationally yet).

The second smallest known planet is also a transiting exoplanet and it was announced recently (GJ 1216b Charbonneau et al. 2009). This object orbits an M-dwarf star. It has a mass of 6.55 M_{Earth} and a radius 2.68 R_{Earth} (Charbonneau et al. 2009). So far, only one star has been found in the habitable zone (Gliese 581d, see Mayor et al. 2009).

2.3. Periods and eccentricities

The periods of the known exoplanets range from 0.79 days (WASP-19b) to 876 years (Fomalhaut b). By this time, no period variation was found in any exoplanets.

The known planets and dwarf planets in our Solar System have a slightly eccentric orbit. Mercury has the most eccentric orbit among the giant planets in our Solar System ($e = 0.205$). Among the dwarf planets we can find small and large eccentricities as well (Ceres has $e = 0.080$ while Eris has $e = 0.442$, Pluto has $e = 0.249$ and Haumea has $e = 0.189$). There is a lower limit for eccentricity (and the natural $e < 1$ upper limit also exists). The closest point of the planetary orbit to the host star is given by

$$q = a\,(1 - e) \tag{2}$$

where q is the closest star-planet distance. There is no orbit for which $q < R_{star}$ (otherwise the planet would fall in the star). From Kepler's third law we know that $a \sim P^{2/3}$, so there is a lower period limit for every eccentricity (and stellar size). In the region of larger eccentricities, we cannot see very short period planets. It is remarkable that there are planets with very large eccentricities. Many examples are known with $e > 0.1$, a few dozens are known with $e > 0.5$ and three exoplanets have larger eccentricity than 0.9.

The gravitational braking (a general relativity effect) is negligible in exoplanets orbiting a normal star. However, the semimajor axis and the eccentricity continuously decrease due to the tidal interaction with the host star. After a certain time the eccentricity becomes zero. This process is called circularization. Its time-scale is given by

$$\tau_c \sim \frac{4}{63}\sqrt{\frac{a^3}{GM_{star}}}\frac{M_{planet}}{M_{star}}\left(\frac{a}{R_{planet}}\right)^5 \tag{3}$$

where a is the semi-major axis of the planet's orbit, M_{star}, M_{planet} is the mass of the star and the planet, respectively, R_{planet} is the planet's radius, and $Q_p \sim 1/2\varepsilon$, where ε is the angle between the line of centers and the tidal bulge (Rasio et al. 1996). Typical circularization time for a hot Jupiter is about 20 million years only. For a similar object in a larger orbit, let us say $a = 0.4$ AU the circularization period is much longer, it is about typically about 13 Gyr. Since a solar-like star spends about 9-12 Gyrs on the main sequence (depending on its mass and chemical composition) burning hydrogen in its center, we can see that if a normal Jupiter (orbiting the star in a distance of 0.4 AU or even more) had initially an eccentric orbit, then this eccentric orbit will not be destroyed and circularized by the planet-star tidal interaction during the main-sequence lifetime of the host star.

Transiting exoplanets are very promising targets to study the effect of gravitational perturbations. If there is no such or other kind of perturbations, i.e. if we have pure Keplerian-motion, then the consecutive revolutions will occur strictly periodically. But the orbital period and other orbital elements of the exoplanets can/will change due to the gravitational perturbation effects. These changes will lead to change in the transit light curve shape and/or to variations in the orbital period, so the consecutive transit instants will deviate from a linear ephemeris, and the length and the shape of the consecutive transit events will differ from each other. (These are usually called TTV-variations, but some authors separate the effects as TTV (Transit Timing Variation), TDV (Transit Depth Variation) and TDuV (Transit Duration Variation) etc). Therefore there are rooms for such investigation, but the time-scales and amplitudes of this kind of variations are in a very wide range (ranging from $\sim$ 1 second to centuries in the timescales and ~second to weeks in the amplitude of the O-C variations, see Steffen 2006). The perturbing third body can be an exomoon, another exoplanet in the system, or a brown dwarf, normal star or a compact object in a more distant orbit revolving the host star: the number of possible configurations and orbits are infinite. A recent review of the most important papers of the field can be found in Csizmadia et al. (2010). It is interesting to note at this conference that the school of celestial mechanic led by Prof. B. Érdi has also already contributed to the field with important and valuable investigatons (many times in collaboration with the Vienna group led by Prof. R. Dvorak), see e.g. Dvorak & Süli (2002), Sándor, Süli & Érdi (2005), Sándor (2006), Érdi, Sándor & Süli (2007), Érdi et al. (2007), Sándor et al. (2007), Süli, Dvorak, Érdi (2007), Schwarz et al. (2007a,b), Érdi et al. (2009), Schwarz, Süli & Dvorak (2009a, b) Dobos, Nagy & Orgoványi (2010). Maybe one of their most important contributions is "the stability catalogue of the habitable zones in extrasolar planetary systems" (Sándor et al. 2007). The present author with his co-workers investigated the TTVs in the CoRoT-1b system (Csizmadia et al. 2010). One of the most promising systems for TTV-studies is TrES-2. TTVs of TrES-2b were predicted by Freistetter, Süli & Funk (2009). Maybe inclination variation was found by Mislis et al. (2010) in this exoplanetary system, but the claim was questioned by Scuderi et al. (2010). Since the object is a Kepler target, the issue may be clarified soon.

The Hungarian Kepler team at the Konkoly Observatory is also contributed to the field by investigating the detectability of exomoons and exorings by different methods (photometry, transit timing and transit duration variations, applying the Rossiter-McLaughlin effect, see e.g. Simon, Szatmáry & Szabó 2007, Simon, Szabó & Szatmáry 2009).

One of the most succesfull ground-based exoplanet teams is the Hungarian Automated Telescope-team, led by Gáspár Bakos. They have already discovered 13 transiting exoplanets (and one more during RV-follow-up; note that WASP-11b = HAT-P-10b). So far the HAT team has detected the shallowest (4 mmag, Bakos et al. 2008) transit from ground – compare to the 0.4 mmag deep of CoRoT-7b detected from space (Léger et al. 2009), which shows that to discover very small planets around FGK dwarfs we have to go to space.

Acknowledgments
The author thanks the organizers for their warm hospitality during the conference.

References

Bakos G Á, Torres G, Pál A, Hartman J, Kovács G et al. 2008 *ApJ* **710** 1724
Baraffe I, Chabrier G and Barman T 2010 *RPPh* **73** 16901
Barnes R, Raymond S N, Greenberg R, Jackson B and Kaib N A 2010 *ApJ* **709** 95
Charbonneau D, Berta Z K, Irwin J, Burke C J, Christopher J et al. 2009 *Nature* **462** 89
Csizmadia Sz, Renner S, Barge P, Agol E, Aigrain S et al. 2010 *A&A* **510** A94
Deeg H, Moutou C, Erikson A, Csizmadia Sz, Tingley B et al. *Nature* in press
Dobos V, Nagy I and Orgoványi J 2010 Stable and habitable systems with two giant planets *JPCS* in this volume
Dvorak R and Süli Á 2002 *CeMDA* **83** 77
Érdi B, Sándor Zs and Süli Á 2007 *PADEU* **19** 179
Érdi B, Nagy I, Sándor Zs, Süli Á and Fröhlich G 2007 *MNRAS* **381** 33
Érdi B, Forgács-Dajka E, Nagy I and Rajnai R 2009 *CeMDA* **104** 145
Fisher D A, Vogt S S, Marcy G W, Butler R P, Sato P et al. 2007 *ApJ* **669** 1336-44
Freistetter F, Süli Á and Funk B 2009 *AN* **330** 469
Guillot T, Burrows A, Hubbard W. B, Lunine J. I and Saumon D 1996 *ApJ* **459** 35
Lammer H, Odert P, Leitzinger M, Khodachenko M L and Panchenko M 2009 *A&A* **506** 343
Léger A, Rouan D, Schneider J, Barge P, Fridlund M et al. 2009 *A&A* **506** 287
Mayor M, Bonfils X, Forveille T, Delfosse X, Udry S et al. 2009 *A&A* **507** 487
Mislis D, Schröter S, Schmitt J H M M, Cordes O and Reif K 2010 *A&A* **510** A107
Moutou C, Hebrard G, Bouchy F, Eggenberger A, Boisse I et al. 2009 *A&A* **498** L5
Queloz D, Bouchy F, Moutou C, Hatzes A, Hébrard G et al. 2009 *A&A* **506** 303
Rasio F A, Tout C A, Lubow S H and Livio M 1996 *ApJ* **470** 1187
Sándor Zs 2006 *CeMDA* **95** 273
Sándor Zs, Süli Á and Érdi B 2005 *PADEU* **15** 211
Sándor Zs, Süli Á, Érdi B, Pilat-Lohinger E and Dvorak R 2007 *MNRAS* **375** 1495
Santos N C 2008 *NewAR* **52** 154
Schneider J 2010 `http://www.exoplanet.eu`
Schwarz R, Dvorak R, Süli Á and Érdi B 2007a *A&A* **474** 1023
Schwarz R, Dvorak R, Süli Á and Érdi B 2007b *AN* **328** 785
Schwarz R, Süli Á and Dvorak R 2009a *MNRAS* **98** 2085

Schwarz R, Süli Á, Dvorak R and Pilat-Lohinger E 2009b *CeMDA* **104** 209
Scuderi L J, Dittmann J A, Males J R, Green E M and Close L M 2010 *ApJ* submitted
Simon A E, Szatmáry K and Szabó Gy M 2007 *A&A* **470** 727
Simon A E, Szabó Gy M and Szatmáry K 2009 *EM&P* **105** 385
Steffen J 2006 PhD Thesis: Detecting new planets in transiting systems *Preprint* arXiv:astro-ph/0609492v1
Süli Á, Dvorak R and Érdi B 2007 *AN* **328** 781
Swift D, Eggert J, Hicks D, Hamel S, Caspersen K, Schwegler E, Collins G and Ackland G 2010 *Preprint* arXiv:1001.4851v1 [astro-ph.EP]
Udry S and Santos N C 2007 *ARA&A* **45** 397
Valencia D, O'Connell R J and Sasselov D 2006 *Icarus* **181** 545
Wagner F, Sohl F, Rauer H, Hussmann H and Grott M 2009 Interior Structure and Bulk Composition of the CoRoT-7b Exoplanet in: European Planetary Science Congress 2009, held 14-18 September in Potsdam, Germany `http://meetings.copernicus.org/epsc2009` p 630

Planet Formation - Implication of Statistical Properties of Exoplanets

Áron Süli

Eötvös Loránd University, Department of Astronomy, Pázmány Péter sétány 1/A, Budapest, H-1518, Hungary

E-mail: a.suli@astro.elte.hu

Abstract. Models of planetary formation are developed based on the observation of our Solar System, star-forming regions and circumstellar disks and on an the ever increasing number of exoplanetary systems. The solar nebula theory and the planetesimal hypothesis are discussed. The latter is found to provide a viable theory of the growth of the terrestrial planets, the cores of the giant planets, and the smaller bodies present in the Solar System. The formation of solid bodies of planetary size should be a common event, at least around young stars which do not have binary companions orbiting at planetary distances. Stochastic impacts of large bodies provide sufficient angular momentum to produce the obliquities of the planets. The masses and bulk compositions of the planets can be understood in a gross sense as resulting from planetary growth within a disk whose temperature and surface density decreased with distance from the growing sun.

1. Introduction

The origin of the Solar System is one of the most intriguing problem of science. For more than two centuries, scientific ideas about how the planets came to be were based almost entirely upon theory. There were few constraints on such speculations: astronomers knew that stars rotated and that the Sun's planet are in almost circular orbits. But beyond these meager knowns, those who speculated on the Solar System's origin had nowhere to turn except to their own fantasy of how a gigantic amount of cosmic matter might have organized itself into a Sun and planetary system. Astronomers, using the laws of physics as then understood, offered variations on the same answer: the Sun and the planets were born form a rotating disk of cosmic gas and dust. The flattened form of the disk constrained the planets that formed from it to have orbits lying in more or less the same plane, all moving in the same direction the disk had turned. This hypothetical disk, the *solar nebula*, is where any discussion of the origin of our Solar System must begin.

The idea of the solar nebula was first proposed by Emanuel Swedenborg (1734), and further developed in the course of the 18th century by the Prussian philosopher and physicist Immanuel Kant (1755). Although his treament of the problem was only qualitative, its precepts were remarkably similar to those considered fundamental today. Kant pictured an early universe evenly filled with thin gas. He thought such a configuartion would have been gravitationally unstable, so it must have drawn itself together into many large dense clumps of gas. Kant correctly assumed that these clumps of gas were rotating and also recognized the importance of

Journal of Physics: Conference Series **218** (2010) 012004 doi:10.1088/1742-6596/218/1/012004

this rotation: as they shrank the rotation spun them out into flattened disks. From one of these disks was our Solar System forged.

The first indirect evidence for disks came from studies of *T Tauri stars*, which are similar in mass to our Sun but very young - roughly a million years old. In the 1980s astronomers realized that about a third of T Tauris stars have "infrared excesses", that is, the amount of infrared radiation they emit is too great to be consistent with their output at visible wavelengths. This can be understood if the stars in question are surronded by halos of dust kept warm by short-wavelength radiation from the stars; the dust then reradiates the energy it receives at longer (infrared and radio) wavelengths. However, the strong infrared signatures implied the presence of enough dust, if distributed evenly in a spehere, to completely block our view of such a star at visible wavelengths. Only if the dust were arrenged in a flattened disk, tilted somewhat to our line of sight, could we expect to see the star itself.

The observational evidence of disks around young stars that were first directly observed in the mid 1980s, initially in the form of dusty debris disks such as those seen around Vega (Aumann et al., 1984) and β Pic (Smith & Terrile, 1984), and later in the form of the gas-rich disks that we now term "protoplanetary" (Sargent & Beckwith, 1987). Advances in telescope technology in the intervening two decades has resulted in the detection, both directly and indirectly, of many more disks around young stars, and these protoplanetary disks, or *proplyds*, are now understood to be commonplace. The Hubble Space Telescope could capture such disks at visible wavelengths in the Orion nebula, an active stellar forming region. The Orion disks are far larger than the Solar System, and the avaialbe material in them are more than enough for planetary systems. The host stars of these disks are very young, at most a few million years old.

Today we realize that Kant, by and large, got it right, and the widely accepted model, the *Solar Nebula Disk Model*, also known as the *Planetesimal Hypothesis* roots in the same fundamentals as Kant's theory. The development of the Solar Nebula Disk Model (hereafter SNDM) has interwoven a variety of scientific disciplines including astronomy, physics, geology, and planetary science. Since the dawn of the space age in the 1950s and the discovery of proplyds and extrasolar planets in the 1990s, the models have been both challenged and refined to account for new observations.

Until the mid 1990s we had confirmed knowledge of only our planetary system and one around a pulsar (Wolszczan & Frail 1992). The first published discovery was made by Campbell et al. (1988), whose radial-velocity observations suggested that a planet orbited the star γ Cep, which is a sub-giant in 57 year long-period binary system. They remained cautious about claiming a true planetary detection, and widespread skepticism persisted in the astronomical community for several years about this and other similar observations. It was mainly because the observations were at the very limits of instrumental capabilities at the time.

The first unambiguous discovery of an exoplanet around a main sequence star 51 Peg was announced on October 6, 1995 by Mayor & Queloz (1995)[1]. In order to better understand the physical processes invloved in the formation of planetary system it is inevitable to observe and study a large sample of extrasolar system. By now this is exactly the situation: since the detection of the planetary companion of 51 Peg the number of exosystems has exceeded 300 providing several constraints and a striking variety of planetary systems for scientist to formulate a general theory for planet formation. On the other hand theoretical studies of planetary formation have become another frontline research in the field of astronomy.

[1] We note that the search for extrasolar planets since 1995 has evolved into a mature field in astrophysics, and in 2002 improved techniques allowed to prove the existance of the planet around the star γ Cep (Hatzes et al. 2003).

Journal of Physics: Conference Series **218** (2010) 012004 doi:10.1088/1742-6596/218/1/012004

1.1. Observations of the Solar System

A complete theory of the origin and evolution of our Solar System and the detected exosystems must explain the observed facts. To begin with I list the major ones of the Solar System:

1. *Orbital inclination, eccentricity and direction:* The inclination of most objects with respect to the Sun's equator is small (coplanar orbits), and they orbit the Sun in the prograde direction (the same sense as the Sun rotates) on nearly circular paths.
2. *Orbital spacing:* The major planets are confined to heliocentric distances of ≤ 30 AU, and the separation between neighboring orbits increases with radial distance. The orbits of the planets do not cross, nor do they approach each other closely. The interplanetary space contains very little debris (aside from the asteroid belt and the Trojans).
3. *Comets (Oort cloud):* Roughly 10^{12} comets of radius $R > 1$ km orbit the Sun at distances beyond 10^4 AU (Oort 1950); this swarm is isotropically distributed with respect to the Sun, apart from asymmetries produced by the tidal field of the Galaxy (Weissman 1990). There is a second comet reservoir, the Kuiper belt - a flattened disk at ≥ 40 AU (Duncan & Quinn 1993a).
4. *Planetary rotation:* Six of the eight major planets rotate around their axis in the same direction in which they revolve around the Sun, and have obliquities (axial tilts) of $\leq 30°$. Venus and Uranus rotate in retrograde direction.
5. *Satellite Systems:* Most planets, aside from Mercury and Venus have natural satellites. Most close-in satellites travel on low inclination, low eccentricity orbits, whereas small distant moons tend to have large inclinations and eccentricities.
6. *Planetary masses:* Together the planets account for $\leq 0.2\%$ of the mass of the Solar System. The most massive planets are Jupiter and Saturn, followed by Neptune and Uranus. The relatively small terrestrial planets orbit closest to the Sun. Between the terrestrial and the giant planets orbit a large number of asteroids. The total mass of the asteroid belt about 1 % that of Mercury.
7. *Angular momentum distribution:* Although the planets represent only a tiny fraction of the total mass, over 98 % of the Solar System's angular momentum is contained in the orbital motions of the Jovian planets. We note, that in contrast the orbital angular momenta of the satellite systems of the giant planets are far less than the spin angular momenta of the planets themselves.
8. *Bulk composition:* The terrestrial planets are composed primarily of dense, refractory (high condensation temperature) rocky material. The low densities of the largest planets imply H and He are the dominant components, yet these planets must still be enriched in high-Z elements (heavier than He) relative to the Sun or their densities would be even less. Jupiter is $\approx 90\%$ H and He by mass and Saturn is $\approx 77\%$ (Hubbard & Marley 1989). Orbiting farther from the Sun, Uranus and Neptune have intermediate uncompressed densities (the density a planet would have if its gravity did not compress it), implying compositions dominated by high-Z material, with $\approx 5-20\%$ H and He (the H and He ratios in these planets are difficult to estimate, as it is not known whether heavier material is primarily very dense rock or lower density melted "ices" of H_2O, NH_3 and CH_4). The moons of the outer planets consits of rock and ice, in varying proportions. Asteroids are rocky, but their low uncompressed densities, relative to the terrestrial planets (Landgraf 1988), suggest some lighter component, perhaps hydrated silicates. Comet Halley consits of a mixture of ice, rock and organics-rich CHON particles. Aside from extremly volatile elements such as H, N, and the noble gases, Halley's elemental abundances appear to be nearly solar.
9. *Asteroid belt:* Roughly 3×10^{24}g of material is spread among countless minor planets orbiting between Mars and Jupiter. The size distribution of asteroids is approximately

a power law with an incremental size index between -2.5 and -3, which has a slope similar to that expected from appropriately adjusted theoretical (Dohnanyi 1969) and experimental (Capaccioni et al. 1986) studies of fragmentation. The eccentricities and inclinations of asteroidal orbits are approximately Rayleigh distributed, with means of 0.14 and 15° respectively. These orbital parameters correspond to random velocities of ≈ 5 km/s, far above the 1.7 km/s escape velocity from 1 Ceres, the largest asteroid.

The SNDM explains several of the above listed observational facts. The nearly circular and cooplanar orbits of Solar System planets strongly support a planetary formation in a flattened circumsolar disk. Star formation models suggest that such disks are a neutral byproduct of a gravitationally collapsing and rotating cloud (Terebey et al. 1984). However, many hydrodynamical calculations have attempted to model stellar collapse and its aftermath, but a completly realistic simulation remains an elusive goal. The low angular momentum material of the cloud formed the Sun, while the high angular momentum material ended up in a flattened disk around the Sun. There is no reason to suppose a different composition for the disk and the Sun: the relative abundance in the elements in the nebula is very similar to that of the Sun.

Proplyds contain a mixture of two components: a gas and condensed matter or dust component. Interstellar grains (possibly formed in the nucleosynthesis of former generations of stars) and nebula condensates make up the the solid component of the disk. Analysis of interplanetary dust and primitive meteorites suggest grain sizes range from $0.05 - 100\mu$m, i.e. roughly 0.1 micron in average (Kerridge & Anders 1988). For the sake of comparison, I note that, the radii of isolated neutral atoms range between 30 and 300 pm (10^{-12} m), so average grains are few 1,000 atoms wide.

To give a lower estimation on the mass of proplyds the *Minimum Solar Mass Nebula* approximation is used: the material in each planet of the Solar System is amended with volatiles to reach the solar composition and the resulted material is then spread out over an annulus centered at the orbit of the planet. The resulting disk would have a mass of $0.01 - 0.02M_\odot$ and a surface density profile which decreases as $\sim r^{-3/2}$, apart from a hole in the region of Mars and the asteroid belt (Weidenschilling 1977a). The bulk of this mass is in the form of gas. Current SNDMs suggest that the mass of the proplyd was several times larger than the Minimum Solar Mass Nebula.

2. Stages of Planet Formation

The process of planetary growth is generally divided for convenience into several distinct stages. The currently accepted sequence of planetary formation is discused in the following sections.

2.1. Early Stage: From Dust to Planetesimals

In what follows I discuss how dust grains grow from sub-μm sizes to $\sim$ km-size bodies. The collapse of a fragment of a molecular cloud to Solar System dimension releases a substantial amount of gravitational potential energy. The bulk of this energy is radiated away during the collapse phase, a significant amount is retained as heat. When infall slows and eventually halts, the disk begins to cool.

2.1.1. Condensation of chemical compounds As the tempereture T decreases various chemical compounds condense into microscopic grains ($\leq 10^{-6}$ m). In the case of the solar disk, the first condensates are silicates (e.g. olivine, $[SiO_4]^{4-}$, pyroxene group $[Si_nO_{3n}]^{2n-}$ etc.). At lower tempereture, typically in the outer region of the proplyd, large quantities of water (H_2O), ammonia (NH_3), methane (CH_4) and other ices can condense (for details see Barshay & Lewis, 1976). Preexisting condensates from the interstellar matter may also be present. Growth of particles then proceeds by collision.

Journal of Physics: Conference Series **218** (2010) 012004 doi:10.1088/1742-6596/218/1/012004

2.1.2. Mutual collision of compounds Observations of dusty disks around T-Tauri stars imply that $\geq 1\%$ of the condensed matter remain in dust phase, which may indicate that collision of particles do not by all means lead to agglomeration of the solid particles. Alternatively it may result from the disruptive collisions between planetesimals, when large amount of dust are ejected.

The mechanical and chemical processes determining dust growth are poorly understood. Experimetal data (Forrest & Witten 1979) and numerical models suggest that weakly bounded particles with fractal geometry may be formed. However, the most primitive meteorites, which are though to have formed in this phase of formation contain chondrules[2]. The large massratio of chondrules (in some cases $\approx 70\%$) in meteorites implies that a significant amount of the fuzzy solid bodies were rapidly heated and cooled prior to being incorporated into larger bodies. However, the environmental setting, the energy source for the heating, and the precursor material are not known (for details of various models of chondrules formation see Morfill et al. 1993).

The motions of small grains in a proplyd are strongly influenced by the gas. The coupling between the gas and the solids with diameter ≤ 1 cm is well described by Epstein's drag law, whereas larger bodies are subject to Stokes drag (Adachi et al. 1976).

The vertical component of the star's gravity causes dust to fall toward the midplane of the proplyd (this process is also called sedimentation). In gas with density $\rho_{\rm g}$ and sound speed c a grain with radius R and density ρ at a distance z from the midplane has a settling velocity $v_{\rm z}$ of

$$\frac{dv_{\rm z}}{dt} = -\frac{\rho_{\rm g}c}{R\rho}v_{\rm z} - \Omega^2 z, \tag{1}$$

where Ω is the orbital frequency. The equilibrium settling rate is given by:

$$v_{\rm z} = -\Omega^2 z\frac{R\rho}{\rho_{\rm g}c}. \tag{2}$$

Grains' settling velocity is lineraly proportional to their radius.

At 1 AU from the Sun, the temperetaure of the proplyd was $\sim 500 - 800$ K and the gas density $\rho_g \sim 10^{-9}$ g/cm^3. Eq. (2) yields e-folding sedimentation times for 1μm dust grains of $\sim 10^6$ years under these conditions, and several e-foldings are required to produce a thin layer in which the dust density and the gas density is comparable. Since dust grains have different radius they approaching the midplane with different velocities: they collide and agglomerate which further increases their settling velocity. This mechanism shortens the sedimentation time by ≈ 2 orders of magnitude. Recent models suggest that the majority of solid material was able to agglomerate into macroscopic size, i.e. ≥ 0.1 mm bodies, and most of these bodies were located to a relatively thin subdisk about the midplane of the proplyd in which the density of solid component was comparable to, or exceeded that of the gas (Weidenschilling & Cuzzi 1993).

The gaseous component of the proplyd is partially supported against the stellar gravity by a pressure gradient in the radial direction, so gas rotates around the star slower than the local Keplerian velocity. The "effective gravity" felt by the gas is

$$g_{\rm eff} = \frac{k^2 M_*}{r^2} - \left(\frac{1}{\rho_{\rm g}}\right)\frac{dP}{dr}, \tag{3}$$

where M_* is the mass of the star. For estimated protoplanetary disk parameters, the gas rotates roughly 0.5% slower than the local Keplerian speed (Adachi et al. 1976).

Larger particles (≥ 10 cm) moving at nearly Keplerian speed thus encounter a headwind, which removes part of their angular momentum and causes them to spiral towards the star.

[2] Chondrules are formed by a rapid heating (within minutes or less) to temperatures between 1500°C and 1900°C and subsequent melting. This is followed by a cooling within one to several hours

Inward drift is greatest for mid-sized particles, which have large surface area to mass ratios and orbit with approximately Keplerian velocities. This inward migration effect is hardly relevant for (*i*) small particles ($\lesssim$ 1 mm), which are so strongly coupled to the gas that the headwind they experience is very weak and for (*ii*) large bodies ($\gtrsim$ 1 km) whose angular momentum is much larger than that of the gas they collide during one revolution. Peak rates for inward drift occur for particles that collide with roughly their own mass of gas in one orbital period. These bodies have a radius of few meters and drift in the terrestrial region of the solar nebula at a rate of up to $\sim 10^6$ km/yr, i.e. a body with mass in the range between 10^2 g and 10^8 g has a decay time at 1 AU as short as 10^2 or 10^3 years! Thus, the material that survies to form the planets must complete the transition from cm size to km size rather quickly.

Several hypotheses have been proposed to overcome this difficulty. A popular set of mechanisms for the formation of planetesimals are rooted in the so-called Goldreich-Ward instability (Goldreich & Ward, 1973). In this scenario the vertical settling of small dust grains steadily increases the dust-to-gas ratio at the midplane, until eventually the thin dust sub-disk becomes gravitationally unstable and fragments. We can estimate the conditions at which this occurs by a simple application of the Toomre (1964) criterion, setting

$$Q = \frac{\sigma\Omega}{\pi G \Sigma_{\rm dust}} = 1, \tag{4}$$

where σ is the velocity dispersion in the dust layer and $\Sigma_{\rm dust}$ is the surface density of the dust layer. If we assume that $\Sigma_{\rm dust} = \Sigma_{\rm gas}/100 \approx 10$ gcm^{-2} and consider a 1 M$_\odot$ star, we see that a velocity dispersion of $\sigma \leq 10$ cms^{-1} is required in order for gravitational instability to fragment the dust layer. Given that the gas sound speed is typically 1 kms^{-1}, we see that the dust sub-layer must be very thin indeed in order to become unstable. However, this suggestion is attractive because it allows for the formation of planetesimals directly from small ($R \lesssim$ 1mm) particles, and thus bypasses the size ranges what are most susceptible to radial drift. If the disk is turbulent, gravitational instabilities are suppressed because the dusty layer remains too thick.

Unfortunately it can be easily shown that, in its simplest form, the Goldreich-Ward mechanism does not result in planetesimal formation. We have seen above that the instability requires that the dust layer be approximately 10^4 times thinner than the gas disk, and consequently at the disk midplane the local dust density dramatically exceeds the local gas density (by a factor ≈ 100). In this layer we therefore expect the solid material to dominate the dynamics, so gas pressure forces are negligible and the dust and gas both orbit at the Kepler speed. Above this layer, however, the gas is significantly sub-Keplerian due to gas pressure, and there is therefore a large velocity shear in the vertical direction. This shear is Kelvin-Helmholz unstable, and the resulting turbulence prevents σ from becoming small enough for instability to set in (Cuzzi et al., 1993).

This problem is the "crux" of the SNDM: How to form planetesimals remains the biggest challenge for modern research in planet formation, and many of the details of this problem still elude us. More recently, several authors have proposed different mechanisms (e.g. aggregation of dust can occur within turbulent eddies) that may provide solutions to the *meter-size problem*. Once this meter-size barrier is passed, continued growth via binary accretion leads to the formation the so called *planetesimals*. In what follows it is supposed that at the end of the early stage a large number of planetesimals orbit the Sun. The settling time of the disk dictates the formation timescale for planetesimals. Most models predict a few 10^4 years.

2.2. Middle Stage: From Planetesimals to Protoplanets

The star's gravity is the dominant force upon planetesimals. The largest perturbation to the orbit of a planetesimal is the gravitational interaction with other solid bodies. The most important nongravitational forces experienced by a planetesimal is the mutual inelastic collision (kinetic energy is not conserved) and gas drag. Gravitational torques may lead to significant orbital evolution of bodies larger than a typical planetesimal.

2.2.1. Velocity distribution The distribution of planetesimal velocities play a crucial rôle in planetery growth. The velocity distribution is modified by gravitational interactions and collisions, which convert the ordered motion with Keplerian circular velocity, $v_{\rm K} = k(M_*)^{1/2}$ to random motion. Due to gas drag larger than $v_{\rm K}$ velocities are decreased, while slower are increased to Keplerian velocity, i.e. gas drag damps the eccentricities and inclinations. In the framework of the kinetic theory of gases the simplest analytic approach for calculating the evolution of planetesimal velocities uses a *particle-in-a-box* approximation, in which the evolution of the mean square planetesimal velocities is calculated. Since in this phase the number of planetesimals which are participating in planet formation process is $\geq 10^9$, only a statistical approach provides a viable method for treating the planetesimal growth.

In the particle-in-a-box approximation the distribution of orbital eccentricities e and inclinations i are described by Rayleigh distribution:

$$f(e,i) = 4\frac{\sigma}{m}\frac{ei}{\langle e^2\rangle\langle i^2\rangle}\exp\left(-\frac{e^2}{\langle e^2\rangle}-\frac{i^2}{\langle i^2\rangle}\right), \tag{5}$$

where m is the mass of the individual planetesimals and σ is the surface mass density of planetesimals with a specific semimajor axis, $\langle e^2\rangle$ and $\langle i^2\rangle$ are the mean square eccentricity and inclination. Direct N-body calculations by Ida & Makino (1992a) have verified the form of Eq. (5). The other angular orbital elements have a uniform distribution within their range.

The distribution of random velocities can be derived from Eq. (5) using the *epicycle approximation*, that is small eccentricities and inclinations are assumed. The result, which is locally (at a specific heliocentric distance r) equivalent to Eq. (5) is a triaxial Gaussian distribution in cylindrical coordinates:

$$f_0(z,\mathbf{v}) = \frac{\Omega\sigma}{2\pi^2c_r^2c_z^2m}\exp\left[-\frac{v_r^2+4v_\vartheta^2}{2c_r^2}-\frac{v_z^2+\Omega^2z^2}{2c_z^2}\right], \tag{6}$$

where

$$2c_r^2 = \langle e^2\rangle v_{\rm K}^2, \quad 2c_z^2 = \langle i^2\rangle v_{\rm K}^2, \tag{7}$$

v_ϑ is the azimuthal component of the velocity relative to the local circular Keplerian velocity, and $\Omega = v_{\rm K}/r$ is the orbital frequency.

The $f_0(z,\mathbf{v})$ distribution is modified by pairwise collisions, gravitational scattering and gas drag. Collisions dissipate some or all of the relative kinetic energy of the colliding bodies while gravitational scattering conserves it but changes the direction of the velocity vectors. Elastic collisions are generally modeled with a Boltzmann collision operator. To account for (i) inelastic collisions and (ii) the gravitational enhancement of collision cross-section the Boltzmann collision operator has been modified; for a detailed discussion see Ida & Nakazawa (1990) and Greenzweig & Lissauer (1990, 1992). The evolution of the distribution function $f_0(z,\mathbf{v})$ can be described by the collisional Boltzmann equation,

$$\frac{\partial f_0}{\partial t} + \mathbf{v}\cdot\nabla_r f_0 - \left[\frac{k^2M_\odot\mathbf{r}}{r^3} + (\mathbf{v}+\mathbf{v}_{\rm K})\cdot\nabla\mathbf{v}_{\rm K}\right]\cdot\nabla_v f_0 = \left.\frac{df_0}{dt}\right|_{\rm coll}, \tag{8}$$

where the velocity is splitted into a Keplerian and a random component. The Boltzmann equation is typically solved by assuming a form for f_0; in the present case it is given by 6.

Gravitational scattering between planetesimals can be modeled with a Fokker-Planck operator similar to that used in stellar dynamics. Both scattering and collision can superimpose random motion on the initially ordered differential rotation of the planetesimals, i.e. these processes perturb the Keplerian shear. Inelastic collisions convert some of the energy of the random motion into heat. In addition to generating random motions, gravitational scatterings also tend to equipartition kinetic energy of random motion among planetesimals of different masses (Stewart & Kaula 1980). This process, which is often referred to as *dynamical friction*, can produce a velocity distribution in which the small particles have a much greater velocity dispersion than do the large ones (Ida & Makino 1992b).

Gas drag damps the orbital eccentricities and inclinations, especially those of smaller bodies. Interested readers are advised to consult Stewart & Wetherill (1988) for the mathematical formulas used to study the evolution of planetesimal velocities.

2.2.2. Runaway growth While gas is still present in the disk eccentricities and inclinations of planetesimals are damped due to aerodynamic gas drag. This counteracts the net effect of collision and scattering and as a result velocities remain low. The simplest accretion picture was first quantified by Safronov (1969). Consider a sphere of radius R moving with velocity v through a uniform medium of density ρ. Then the accretion rate is

$$\frac{dm}{dt} = \pi R^2 \rho v. \tag{9}$$

This rate is the geometrical accretion limit, and $\dot{m} \sim m^{2/3}$. However, the gravity of the body plays an important rôle by gravitationally enhancing the cross-section of the body. This effect was first recognized by Safronov: bodies that are larger than the typical size can accelerate their growth rate due to *gravitational focusing*, i.e. gravitationally enhancing the cross-section of the body (Safronov 1969):

$$\frac{dM}{dt} = \pi R^2 \left[1 + \left(\frac{v_{\rm esc}}{v_{\rm rand}}\right)^2\right], \tag{10}$$

where $v_{\rm esc} = k\sqrt{2M/R}$ is the escape velocity from the body's surface, and $v_{\rm rand}$ represents the velocity dispersion of planetesimals. The second term in the parentheses represent the gravitational enhancement of the accretion cross-section. The quantity $2\theta = v_{\rm esc}/v_{\rm rand}$ is known as the *Safronov number*, and $1 + 2\theta$ is referred to as the *gravitational enhancement factor*.

While random velocities are small, gravitational focusing can increase the growth rates of bodies by a factor of hundreds, such that $\dot{m} \sim m^{4/3}$, leading to a phase of *runaway growth* (Ida & Makino, 1992a). The term *runaway growth* or *runaway accretion* refers to an evolutionary path where the largest planetesimal in the local region grows much more rapidly than the remainder of the population and therefore becomes detached from the continuous size distribution. These bodies are the progenitors of the *protoplanets*. The length of this phase depends on the timescale for random velocities of planetesimals to reach the escape speed of the larger bodies. For smaller ($\lesssim$ 100 m-sized) planetesimals, gas drag is more efficient such that runaway growth can sustain for a longer time and planetery bodies may be larger.

The runaway growth phenomenon has been seen in numerical simulations (cf. Richardson et al. 2000, Kokubo & Ida 2000). Their results show that the most massive bodies have significantly smaller eccentricities than the remainder of the population, presumably they have been damped by dynamical friction. With runaway accretion the timescale for growth reduce to $\leq 10^5$ years. The simulations indicate that the typical separation between protoplanets formed in this phase is $\approx 4 - 10$ mutual Hill-radii, which is defined as

$$R_{H,m} = \frac{1}{2}(a_1 + a_2)\left(\frac{m_1 + m_2}{3M_*}\right)^{1/3}, \tag{11}$$

where a_1 and m_1 denote the semimajor axis and mass of body 1, etc. (Kokubo & Ida 1996).

The mass of the protoplanets depends on the assumption of planetesimal sizes. In a swarm of $m = 10^{23} - 10^{24}$g (radius 200 - 400 km) planetesimals, Ida & Makino (1993) calculated $\sim 10^{-3} - 10^{-2} M_\oplus$, where the mass of the Earth, $M_\oplus \approx 6 \times 10^{27}$g. Considering a perhaps more realistic population of planetesimals with mass $\approx 10^{19}$g, for the protoplanets $\sim 10^{-5} - 10^{-6} M_\oplus$ are obtained. Thus, although runaway accretion is much more rapid than other modes of accretion (e.g. geometric or oligarchic), it ceases long before protoplanets approaching an Earth mass can form.

2.2.3. Oligarchic growth As it was discussed above the initial growth mode in a disk of accreting planetesimals is runaway growth, where the mass doubling time for the largest bodies is the shortest. Runaway growth allows relatively short formation times and it is followed by a more lengthy phase of oligarchic growth mode when the largest bodies are still orders of magnitude below an Earth mass. The timescale of oligarchic growth dominates over that of runaway growth.

When these runaway bodies, or protoplanets, become sufficiently massive, it is their gravitational scattering, which is often called *viscous stirring* which dominates the random velocity evolution of the background planetesimals, rather than the interactions among the planetesimals. Since the accretion cross-section of a protoplanet is smaller among planetesimals with higher random velocities, protoplanet growth now switches to a slower, self-limiting mode, in which the mass ratio of any two protoplanets at adjacent locations in the disk approaches unity over time. Ida & Makino (1993) investigated this transition analytically and through N-body simulations, and Kokubo & Ida (1998, 2000, 2002) studied the subsequent accretion mode, giving it the name *oligarchic growth.* In this phase just a few dozens of larger bodies, the "oligarchs" dominate the dynamics of the system.

The oligarhic growth phase begins when the dynamical friction of the planetesimals is insufficient to keep the eccentricities and inclinations of the protoplanets very low. In such a configuration protoplanets are not sufficiently isolated from one another to be dynamically stable for a long period of time. Mutual gravitational scattering eventually pump up the relative velocities of the protoplanets which decreases the collisional cross-sections and hence increases the accumulation time. During the oligarhic growth the protoplanets clear their feeding zone by either agglomerating the remaining planetesimals or ejecting them out of the zone.

The final mass of the protoplanets can be approximated by the mass available within the feeding zone. For the case of a protoplanet on a circular orbit the standard theory of the restricted three body problem places an upper bound on the initial semimajor axis separation b_H, that may lead to collision. It is convenient to introduce Hill-scaled units for planetesimal with mass m, eccentricity e, inclination i, separation in semimajor axis a from the protoplanet:

$$e_H \equiv \frac{ea}{h}, \quad i_H \equiv \frac{ia}{h}, \quad b_H \equiv \frac{\Delta a}{h}, \tag{12}$$

where $h = (m/3M_*)^{1/3}a$ is the Hill sphere of the protoplanet and M_* is the mass of the star in solar units. Neglecting gas drag and interactions with other planetesimals, a planetesimal whose orbital elements satisfy the inequality:

$$\frac{3}{4}b_H^2 - e_H^2 - i_H^2 \geq 9 \tag{13}$$

cannot enter the protoplanet's Hill sphere. For $e_H = 0$ and $i_H = 0$ this inequality gives $b_H \geq 2\sqrt{3} \approx 3.46$. Planetesimals on circular orbit with zero inclination and initial separation b_H less than $2\sqrt{3}$ will suffer a close encounter and eventually merge with the protoplanet. Thus the accretion zone, often referred to as the *feeding zone* embedded in a disk of low random velocity extends over the region

$$b_H \leq B, \tag{14}$$

where B depends on the magnitude of other perturbations on the planetesimals, and typically is $\sim 3.5 - 4$ in a quiescent disk.

The mass of a protoplanet, assuming that it revolves at astrocentric distance r and has accreted all the solid material within an annulus of width $2\Delta r$ is

$$m = \int_{r-\Delta r}^{r-\Delta r} 2\pi r' \sigma(r') dr' \approx 4\pi r \Delta r \sigma(r). \tag{15}$$

Setting $\Delta r = Bh = Br(m/3M_*)^{1/3}$, one obtains from Eq. (15) the *isolation mass* to which a protoplanet during the oligarhic growth phase may grow:

$$m = \frac{(4\pi B r^2 \sigma)^{3/2}}{(3M_*)^{1/2}} = 2.1 \times 10^{-3} \left(\frac{Br^2\sigma}{2\sqrt{3}}\right) \left(\frac{M_\odot}{M_*}\right) M_\oplus \tag{16}$$

For example, assuming $B = 2\sqrt{3}$, a minimum mass solar nebula with $\sigma = 10$ gcm^{-2} at 1 AU implies protoplanet isolation at 0.066 $M_\oplus$; whereas $\sigma = 3$ gcm^{-2} at 5 AU implies protoplanet isolation at 1.36 $M_\oplus$.

Therefore at the end of this phase only such protoplanets are formed, whose masses are a few perecent of the Earth mass in the terrestrial region, and roughly an Earth mass at Jupiter. One possibility to get larger mass is to start with more than the minimum mass solar nebula. In this case the excess material must be cleared away. This can easily happen in the vicinity of Jupiter as that planet is large enough to eject bodies, but in the terrestrial region, where material resides deep in the potential well of the Sun, it is not evident how the extra material would be lost.

Two plausible mechanisms may contribute to continue the growth of the protoplanets once the isolation mass has been reached:

1. The diffusion of more material into the feeding zone from neighboring regions via scattering between planetesimals, perturbations by close protoplanets and gas drag.
2. Radial motion of a protoplanet may bring it into zones not depleted of planetesimals or enable a low velocity collision and subsequent merge with a neighboring protoplanet.

It is important to realize that terrestrial protoplanets are the same objects as giant planet cores, assuming giant planets to form via the buttom-up core-accretion scenario.

2.3. Late Stage: From Protoplanets to Terrestrial Planets

At the end of the middle stage we have a regular distribution in a of the protoplanets with roughly constant mass. There is also a swarm of smaller planetesimals, which may or may not be relevant. At the beginning of the late stage a.k.a *giant impact* phase, most of the mass is contained in protoplanets, so dynamical friction is no longer playing an important rôle in their orbital evolution. Even if protoplanets form in circular orbits, mutual gravitational perturbations among several bodies can induce eccentricities of ~ 0.01, which is sufficient to enable their orbits to cross so the bodies can suffer close encounters. These are the initial conditions for the last stage of planet formation.

In this stage there are to few bodies to use statistical methods, therefore to track the evolution of the system one must use direct N-body integration methods. Once the protoplanets have pertubed one another into crossing orbits, their subsequent orbital evolution is determined by close encounters and collisions. Almost all simulations performed to date assume that the collisions are inelastic and all the material in the coliding bodies end up in the resulting body of the collision. However, giant collisions between protoplanets may not always be accretionary: high-speed or off center collisions can actually erode the body.

Journal of Physics: Conference Series **218** (2010) 012004 doi:10.1088/1742-6596/218/1/012004

For dynamical environments typical of late-stage accretion models Agnor & Asphaug (2004) estimated that more than half of all collisions between same-sized planetary embryos do not result in accumulation into a larger body. While their initial results are limited to a single mass ratio, they suggest that nonaccretionary collisions are typical during the end of terrestrial planet formation. The collisional and dynamical accretion of planets are coupled. For example, the reduced accretion efficiency appears to lengthen the timescale of planet formation by a factor of 2 or more, relative to perfect mergers. The production of significant erosional debris, however, might alter the dynamical environment in ways that remain largely unexplored - for instance, damping the system to lower relative velocities.

Benz et al. (1988) simulated giant collision between a proto-Mercury and a planet one-sixth its size. According to the results it would lead to the loss of most of the silicate mantle of Mercury and thereby account for its anomalously high density. Other simulations in the field lend support to the giant impact hypothesis for the origin of Earth's Moon; during the final stage of accumulation, an Earth-size planet is typically happen to collide with several objects as large as the Moon and often as massive as Mars. The obliquities of the rotation axes of the planets also provide evidence of the giant impacts during this epoch.

This stage of formation was pioneered by Chambers & Wetherill (1998), who simulated the terrestrial-planet formation using N-body integrations, in three dimensions, of disks of up to 56 initially isolated, nearly coplanar planetary embryos, plus Jupiter and Saturn. Gravitational perturbations between embryos until their orbits become crossing, allowing collisions to occur. Further interactions produce large-amplitude oscillations in e and the i and prevent objects from becoming re-isolated during the simulations. The largest objects tend to maintain smaller e and i than low-mass bodies, suggesting some equipartition of random orbital energy, but accretion proceeds by orderly growth. The simulations typically produce two large planets interior to 2 AU, whose time-averaged e and i are significantly larger than Earth and Venus. The accretion rate falls off rapidly with heliocentric distance, and embryos in the "Mars zone" (1.2 ¡ a ¡ 2 AU) are usually scattered inward and accreted by "Earth" or "Venus," or scattered outward and removed by resonances, before they can accrete one another. The asteroid belt (a ¿ 2 AU) is efficiently cleared as objects scatter one another into resonances, where they are lost via encounters with Jupiter or collisions with the Sun, leaving, at most, one surviving object. Accretional evolution is complete after 3×10^8 years in all simulations that include Jupiter and Saturn. The number and spacing of the final planets, in the simulations, is determined by the embryos' eccentricities, and the amplitude of secular oscillations in e, prior to the last few collision events.

A large number of simulations were conducted in recent years exploring a wide range of initial conditions. In general the end result of all these simulations is the formation a few (2 - 5) terrestrial planets on a timescale of about 10^8 years. An important feature is that planetesimal orbits execute a random walk in semimajor axis due to successive gravitational encounters. The resulting widespread mixing of material throughout the terrestrial planet region greatly diminishes any chemical gradients that may have existed in the early stages of planetesimal formation.

The above described planet formation mechanism is unlikely to be as purely sequential as layed out. Grain growth and even the accretion of large planetesimals may well begin during the epoch when a proplyd is still accreting and redistributing material. Given that the theories of each of these epochs are still rather primitive, a sequential study of each stage is probably adequate. However, to the extent that planetary growth depends on, e.g. the initial size distribution of planetesimals, one must recognize that various processes currently being treated as separate events occur simultaneously and may effect one another.

3. Considerations For SNDM From Statistical Properties Of Exoplanets

During the past decade, our understanding of planetary formation had to integrate several new peculiar characteristics. With the number of detected planets regularly increasing, one must continuously reexamine the statistical properties of the derived orbital elements and stellar-host characteristics, and search for constraints for the different planet formation and evolution scenarios. Some confidence to observed trends is brought by the regularly increasing number of candidate detections. The most stunning feature of the sample is the variety of orbital characteristics. This variety challenges the conventional views of planetary formation. The goal now is to interpret the observed orbital distributions in terms of constraints for the planetformation models.

Many studies have discussed from an observational point of view the statistical properties of the extrasolar planets, analyzing various distributions and correlations in order to address the following and several other issues, as reviewed by Udry & Santos (2007).

3.1. Occurrence Rate of Giant Planets

The most direct statistical property of a planet-search program is the fraction of detected planets among the surveyed stars. In the CORALIE planet search sample (1650 FGK stars; Udry et al. 2000), 0.8% of stars have giant planets ($m_2 \sin i \geq 0.2 M_{\rm Jup}$) with separations less than 0.1 AU (hot Jupiters), and $63/1120 = 5.6\%$ of stars have giant planets at separations out to 4 AU. For planets more massive than 0.5 $M_{\rm Jup}$, in the Lick+Keck+AAT sample (1330 FGKM stars; Marcy et al. 2005) it was found that 1.2% of the stars host hot Jupiters and 6.6% of stars have planets within 5 AU. Within error bars, these two large samples are in good agreement. The planet occurrence frequency is better determined as a function of planet mass and orbital period using Monte Carlo simulations to estimate detection incompleteness. This has been done only in a few cases. For the ELODIE program (Perrier et al. 2003), it is estimated for planets more massive than 0.5 $M_{\rm Jup}$ a corrected fraction of $0.7 \pm 0.5\%$ for hot Jupiters with $P \leq 5$ days, and of $7.3 \pm 1.5\%$ for planets with periods smaller than 3900 days (~ 4.8 AU).

3.2. Distribution of planetary masses

Companion minimum mass and separation from the star are the two properties of the system directly obtained from radial-velocity measurements. The known planetary mass distribution, an important constraint for formation models, is discussed in this section separately for giant gaseous planets and for the emerging population of small-mass solid planets.

3.2.1. Giant gaseous planets It was already clear after the detection of a handful of extrasolar planets that these objects could not be considered the low-mass tail of stellar companions in binary systems (with low $m_2 \sin i$ because of nearly face-on orbital inclinations). The strong bimodal aspect of the secondary-mass distribution to solartype primaries has generally been considered the most obvious evidence of different formation mechanisms for stellar binaries and planetary systems (e.g., Udry et al. 2002). No real clear limit exists as the distribution goes continuously into the brown-dwarf regime but with a decreasing number of candidates. The interval between the two populations (planets and stellar binaries), the *brown-dwarf desert*, corresponding to masses between $\sim 15 M_{\rm Jup}$ and $\sim 60 M_{\rm Jup}$, contains only a few objects, at least for orbital periods shorter than a decade. There is, however, a probable overlap of the planet and binary distributions in the 10 - 20 $M_{\rm Jup}$ domain. At this point, it is not easy to differentiate low-mass brown dwarfs from massive planets just from their $m_2 \sin i$ measurements without additional information on the formation and evolution of these systems. Even Hipparcos astrometry has been shown not to reach a sufficient precision to unambiguously probe planetary companions.

Toward the low-mass planets, a clear rise of the distribution is observed. Marcy et al. (2005) proposed a power-law-type function, $dN/dM \propto M^{-1.05}$, to describe the result from their FGKM sample. This fit is not affected by the unknown $\sin i$ distribution, which simply scales in the vertical direction. Below the mass of Saturn, the distribution is strongly affected by the detection bias inherent to radial-velocity measurements. We are nevertheless observing the onset of a new population of very light planets.

3.2.2. A new population of solid planets The low-mass edge of the planet mass distribution is poorly defined because of observational incompleteness. However, although the lowest-mass planets are difficult to detect, in the past years about 40 planets with masses in the Uranus-Neptune range $\leq 20M_{\oplus}$) have been detected. Because of their small masses and locations in the system, close to their parent stars, these light planets may well be composed mainly of a large rocky/icy core (e.g., Brunini & Cionco 2005, Alibert et al. 2006). It is possible that they either lost most of their gaseous atmosphere or simply formed without accumulating a substantial one.

The discovery of very low-mass planets so close to the detection threshold of radial-velocity surveys, and over a short period of time, suggests that this kind of object may be rather common. Moreover, at larger separations, the microlensing technique is finding similar mass objects (the lightest with a minimum mass of $\sim 2.0M_{\oplus}$, Mayor et al. 2009), indicating that smaller-mass planets can be found over a large range of separations. This is in complete agreement with the latest Monte Carlo simulations of accretion-based planet formation models predicting large numbers of solid planets. In this line of thought, we already can note that the detected Neptune-mass planets build a distribution of their own with a gap starting to appear between them and the more common Jupiter-mass planets. Although this result is still strongly observationally biased, it will be interesting to see if the properties of the two populations show further differences, especially when the available statistics increase.

3.3. Orbital Period Distribution of Exoplanets

The distribution of periods of giant exoplanets is characterized by two main features: a peak around 3 days and an increasing distribution with period. The hot Jupiters were completely unexpected before the first exoplanet discoveries. The SNDM (e.g., Mizuno 1980, Pollack et al. 1996) suggests that giant planets form first from ice grains beyond the snowline. Once the solid core is massive enough gas could rapidly accrete (Safronov 1969) over the lifetime of the protoplanetary disk ($\sim 10^7$ year, e.g., Haisch, Lada & Lada 2001). During this process, they are also supposed to undergo a migration episode moving from their birth place closer to the central star (see, e.g., Lin, Bodenheimer & Richardson 1996; Papaloizou & Terquem 2006), where they have to stop before falling onto the star. Several stopping mechanisms have been proposed, invoking, e.g., a magnetospheric central cavity of the accretion disk, tidal interactions with the host stars, Roche-lobe overflow by the young inflated giant planet, or photoevaporation. The question is, however, still debated. Alternatives invoke in situ formation, possibly triggered through disk instabilities (Boss 1997). Even in such cases, subsequent disk-planet interactions leading to migration is expected to take place as soon as the planet has formed. The observed pile up of planets with periods around 3 days is believed to be the result of migration and final stopping mechanism (see, e.g., Udry, Mayor & Santos 2003, and references therein, for a more detailed discussion). The exact distance range related to the stopping mechanism is not well defined. A small tail of the distribution points toward the small separations. In particular, three of the transiting planets in the OGLE survey have periods shorter than 2 days (very hot Jupiters). Such short periods, although easy to detect, are not found in the radial-velocity surveys, suggesting that those objects are about 10 times less numerous than hot Jupiters.

Another interesting feature of the period distribution is the rise of the number of planets with increasing distance from the parent star, up to a separation corresponding to the duration

Journal of Physics: Conference Series **218** (2010) 012004 doi:10.1088/1742-6596/218/1/012004

limit of most of the older surveys (4 - 5 AU). The position of the maximum of the distribution is unknown. However, a conservative flat extrapolation of the present distribution out to larger separations would approximately double the occurrence rate of planets (Marcy et al. 2005). This extrapolation hints that a large population of yet undetected Jupitermass planets may exist between 3 - 20 AU. This is of prime importance for the directimaging projects under development on large telescopes such as SPHERE, the VLT Planet Finder, or the Gemini Planet Imager, and space-based imaging missions such as NASA's *James Webb Space Telescope* (JWST) and *Terrestrial Planet Finder*, or ESA's *Darwin*.

3.3.1. Period distribution of Neptune-mass planets Although the number of known Neptune-mass planets is small, it is interesting to see how their orbital parameters compare with properties of giant extrasolar planets. Because of the tiny radial-velocity amplitude they induce on the primary stars, limiting detections to short orbital periods, a meaningful comparison can only be done for giant planets with periods smaller than $\sim$ 20 days. In the period distribution of Neptune-mass planets no pile up can be observed, it is rather flat up to 30 days. We even have a candidate with a six-month period. This hints at another difference in the properties of Neptune- and Jupiter-mass planets at short periods.

3.4. The Period-Mass Diagram

The orbital period distribution has highlighted the importance of migration processes to explain the observed configuration of planetary systems. When coupling period and mass, further striking features appear in the distribution: the first noticeable characteristic of the distribution is the lack of massive planets on short-period orbits (Pätzold & Rauer 2002, Udry et al. 2002). Ignoring the multiple-star systems, one completely misses candidates with masses larger than $\sim 2M_{\rm Jup}$ and periods smaller than $\sim$ 100 days. This is not an observational bias as these candidates are the easiest ones to detect. Migration scenarios may naturally result in a paucity of close-in massive planets. Type II migration, when a planet is massive enough ($\geq 0.5-1.0M_{\rm Jup}$) to clear a gap in the disk, has been shown to be less effective for massive planets (Nelson et al. 2000); i.e., massive planets are stranded at wider separations than low-mass planets. Moreover, when a migrating planet reaches small separations from the star, some process related to planet-star interactions could promote mass transfer from the planet to the star, decreasing the mass of the migrating planet, or cause massive planets to fall into the central star (Pätzold & Rauer 2002).

Another interesting feature of the distribution is the rise in the maximum planet mass with planet-star separation (Udry, Mayor & Santos 2003). Massive planets are easily detected at small separations, yet they preferentially reside in more distant orbits. This can be understood in the context of the migration scenario as well. More massive planets are expected to form further out in the protoplanetary disk, where raw materials for accretion are available along a longer orbital path, thus providing a larger feeding zone. Then, migration may be more difficult to initiate as a larger portion of the disk has to be disturbed to overcome the inertia of the planet.

To explain the observed correlation, it has also been suggested that multiplanet chaotic interactions preferentially move low-mass (low-inertia) planets either inward or outward in the system, whereas massive (high-inertia) planets are harder to dislodge from their formation site (Weidenschilling & Marzari 1996, Marzari & Weidenschilling 2002). It is however still an open question whether the correct frequency of hot Jupiters can be reproduced this way.

Simulations of migrating planets in viscous disks are consistent with the observation of a decrease in the efficiency of migration with increasing planet mass (Trilling et al. 1998; Nelson et al. 2000). Therefore, it seems reasonable to expect that a large number of massive planets may reside on longperiod orbits, some still undetected because of the time duration of the present surveys.

4. Conclusions

The planetesimal hypothesis provides a viable theory of the growth of the terrestrial planets, the cores of the giant planets, and the smaller bodies in the Solar System. The formation of giant planets, which contains significant amounts of H_2 and He, requires a rapid growth of planetary cores so that gravitational trapping of gas can occur prior to the dispersal of the gas from the proplyd.

The summarized theory of planet formation provides excellent or acceptable explanations of the causes of several of the observed Solar System properties listed in the Introduction, but less complete or satisfactory for several others. Many issues remain to be solved: the details of planetesimal formation are still poorly understand, although recent results suggest that several process including turbulence and migartion of meter-sized bodies acting simultaneously might be the solution. The details and consequences of giant impacts are not well known, in terms of the fate of collisional debris and compositional changes induced by the impacts. Moreover, a number of the effects of the external parameters could be much more decisive in the formation process than we think it now.

An important challenge in the field of observation is that of directly imaging a planet orbiting a solar-type star. The development of a new generation of adaptive-optics systems, promises a great improvement in this field. All these elements will permit us to better understand the mechanisms leading to the formation of planetary systems like our own, and will thus represent an important step toward the search for life in the Universe. Once earth-like planets orbiting in the habitable zone are known, the search for life in these systems will undoubtedly follow. In the very near future, humanity has to prepare itself to find out that the whole Universe may be teeming with life.

References

Adachi I Hayashi C and Nakazawa K 1976 *Prog. Theor. Phys.* **56** 1756
Alibert Y et al 2006 *Astronomy and Astrophysics* **455** L25
Aumann H H et al 1984 *ApJ* **278** L23
Agnor C and Asphaug E 2004 *The Astrophysical Journal* **613** L157
Barshay S S and Lewis J S 1976 *Annu. Rev. Astron. Astrophys.* **14** 81
Benz W, Slattery W L and Cameron A G W 1988 *Icarus* **74** 516
Bodenheimer P and Pollack J B 1986 *Icarus* **67** 391
Boss A P 1997 *Science* **276** 1836
Brunini A and Cionco R G 2005 *Icarus* **177** 264
Campbell B, Walker G A H and Yang S 1988 *Astrophysical Journal* **331** 902
Capaccioni F, Cerroni P, Coradini M, di Martino M and Farinella P 1986 *Icarus* **66** 487
Chambers J E and Wetherill G W 1998 *Icarus* **136** 304
Cuzzi J N, Dobrovolskis A R and Champney J M 1993 *Icarus* **106** 102
Dohnanyi J S 1969 *J. Geophys. Res.* **74** 2531
Duncan M J and Quinn T R 1993a *Annu. Rev. Astron. Astrophys.* **31** 265
Hatzes A P, Cochran W D, Endl M, McArthur B, Paulson D B, Walker G A H, Campbell B and Yang S 2003 *The Astrophysical Journal* **599** 1383
Forrest S R and Witten T A Jr 1979 *J. Phys.*A **12** L109
Goldreich P and Ward W R 1973 *ApJ* **183** 1051
Greeberg R, Hartmann W K, Chapman C R and Wacker J F 1978 *Icarus* **35** 1
Greenzweig Y and Lissauer J J 1990 *Icarus* **87** 40
Greenzweig Y and Lissauer J J 1992 *Icarus* **100** 440
Haisch K E Jr, Lada E A and Lada C J 2001 *The Astrophysical Journal* **553** L153
Hubbard W B and Marley M S 1989 *Icarus* **78** 102
Ida S and Makino J 1992a *Icarus* **96** 107
Ida S and Makino J 1992b *Icarus* **98** 28

Ida S and Makino J 1993 *Icarus* **106** 210
Ida S and Nakazawa K 1990 *Icarus* **86** 561
Kokubo E and Ida S 1996 *Icarus* **123** 180
Kokubo E and Ida S 1998 *Icarus* **131** 171
Kokubo E and Ida S 2000 *Icarus* **143** 15
Kokubo E and Ida S 2002 *Astrophys. J.* **581** 666
Landgraf W 1988 *Astron. Astrophys.* **191** 161
Lin D N C, Bodenheimer P and Richardson D C 1996 *Nature* **380** 606
Lissauer J J 1987 *Icarus* **69** 249
Marcy G, Butler R P, Fischer D, Vogt S, Wright J T, Tinney C G and Jones H R A 2005 *Progress of Theoretical Physics Supplement* **158** 24
Marzari F and Weidenschilling S J 2002 *Icarus* **156** 570
Mayor M et al 2009 *A&A* **507** 487
Mizuno H 1980 *Prog. Theor. Phys. Suppl.* **64** 54
Morfill G, Spruit H and Levy E H 1993 *Protostars and planets III* (A93-42937 17-90), p. 939-978.
Nelson R P, Papaloizou J C B, Masset F and Kley W 2000 *Monthly Notices of the Royal Astronomical Society* **318** 18
Oort J H 1950 *Bull. Astron. Inst. Neth.* **11** 91
Papaloizou J C B and Terquem C 2006 *Reports on Progress in Physics* **69** 119
Pätzold M and Rauer H 2002 *The Astrophysical Journal* **568** L117
Perrier C, Sivan J-P, Naef D, Beuzit J L, Mayor M, Queloz D and Udry S 2003 *Astronomy and Astrophysics* **410** 1039
Pollack J B, Hubickyj O, Bodenheimer P, Lissauer J J, Podolak M and Greenzweig Y 1996 *Icarus* **124** 62
Richardson D C, Quinn T, Stadel J and Lake G 2000 *Icarus* **143** 45
Safronov V D 1969 *Evolution of the Protoplanetary Cloud and Formation of the Earth and Planets* Moscow: Nauka. Engl. Transl. NASA TTf-677 1972
Sargent A I and Beckwith S 1987 *ApJ* **323** 294
Smith B A and Terrile R J 1984 *Science* **226** 1421
Stewart G R and Kaula W M 1980 *Icarus* **44** 154
Stewart G R and Wetheril G W 1988 *Icarus* **74** 542
Terebey S, Shu F H and Cassen P 1984 *Ap. J.* **286** 529
Toomre A 1964 *ApJ* **139** 1217
Trilling D E, Benz W, Guillot T, Lunine J I, Hubbard W B and Burrows A 1998 *Astrophysical Journal* **500** 428
Udry S, Mayor M, Naef D, Pepe F, Queloz D, Santos N C, Burnet M, Confino B and Melo C 2000 *Astronomy and Astrophysics* **356** 590
Udry S, Mayor M, Naef D, Pepe F, Queloz D, Santos N C and Burnet M 2002 *Astronomy and Astrophysics* **390** 267
Udry S, Mayor M and Santos N C 2003 *Astronomy and Astrophysics* **407** 369
Udry S and Santos N C 2007 *ARAA* **45** 397
Weidenschilling S J 1977a *Astrophys. Space Sci.* **51** 153
Weidenschilling S.J 1977b *MNRAS* **180** 57
Weidenschilling S J and Cuzzi J N 1993 Protostars and planets III 1031
Weidenschilling S J and Marzari F 1996 *Nature* **384** 619
Weissman P R 1990 *Nature* **344** 825

On the formation of close binaries – an interferometric look on Algol(s)

Tamás Borkovits[1], **Zsolt Paragi**[2,3] **and Szilárd Csizmadia**[4]

[1] Baja Astronomical Observatory, H-6500 Baja, Szegedi út, Kt. 766, Hungary
[2] Joint Institute for VLBI in Europe, Postbus 2, 7990 AA Dwingeloo, The Netherlands
[3] MTA Research Group for Physical Geodesy and Geodynamics, H-1585 Budapest, P. O. Box 585., Hungary
[4] Institute of Planetary Research, DLR Rutherfordstr. 2. D-12489, Berlin, Germany

E-mail: borko@electra.bajaobs.hu

Abstract. Both theoretical and indirect observational evidences suggest that most of the closest binary stars with orbital periods of few days (as well as hot Jupiter-type exoplanets) formed with an original separation larger by 1-2 orders of magnitude than the present one. Consequently, an orbital shrinking mechanism(s) must be present that extracts the angular momentum from a primordial binary. From the possible mechanisms studied in the literature we describe the so-called 'Kozai cycles combined with tidal friction' (KCTF), which gives some definite predictions for the statistical properties of the orbital elements in such close binaries. Due to the recent powerful observing techniques, such as optical and radio (VLBI) interferometry, we can already peer inside such close binaries, so we have some opportunity to check these predictions. We examine the practical details, constraints, restrictions about the inquiring of the necessary information from specific triple systems by interferometric (as well as polarimetric) observations. Finally, to illustrate this we give a brief description of our recent combined optical and radio interferometric measurements of Algol.

1. Introduction

Ternary stellar systems are common in the Galaxy. If n denotes the number of components in a stellar system then the expected relative frequency of n-body systems is proportional to $1/n!$ (Batten 1973) and therefore the relative frequency of ternary systems compared to the binary ones is about $2!/3! = 1/3$ which shows that we can expect many (mainly) hierarchical triple systems in the sky. The continuously updated catalog of Tokovinin (1997) lists about 1000 such systems, many of them containing very close binaries with a characteristic orbital period of days or even hours.

Although the statistics of triple and multiple stellar systems (i. e. the number of the components, period distributions, mass ratios, orbital properties) are very poorly known, and are far from the completeness even in the case of the brightest systems (c. f. Eggleton and Tokovinin 2008), some systematic surveys suggest that most of the shortest period binaries have at least one additional stellar component. For example, Tokovinin et al. 2006 surveying a representative sample of 165 solar-type spectroscopic binaries concluded that the fraction of the (solar type) spectroscopic binaries with additional companions is about $63\% \pm 5\%$; this ratio reaches 96% for $P < 3$ d. This result has a great importance from the point of view of the stellar or (more strictly) of the binary formation theories. According to our recent knowledge about stellar formation the minimal initial orbital separation of primordial binaries are strongly limited due to the large size of a protostar. For example, a proto-star with solar mass has a radius of $\sim 6R_\odot$ at an age of 7×10^4 yr (D'Antona and Mazzitelli 1994); which limits the minimal period of a marginally contact binary formed by two such proto-stars at $P = 3.4$ d. Thus, the theory strongly suggests that many of the binary stars underwent a very strong orbital shrinkage (even by 1-2 orders of magnitude) during their lifetime. (Note, it is also thought that hot-Jupiters found in several exoplanetary systems should also have been the subject of some orbital shrinking mechanisms.) Different mechanisms are described in the literature. A short list of them, can be found e. g. in Tokovinin et al. (2006) and Fabrycky and Tremaine (2007). In the next section we give a more detailed description about one of them, the so-called 'Kozai cycles combined with tidal friction' (KCTF).

2. Kozai cycles

The Kozai resonance (or recently it is frequently mentioned as Kozai cycle[s]) first was described by Kozai (1962) investigating secular perturbations of asteroids. The first (theoretical) study of this phenomenon with respect to multiple stellar systems can be found in the work of Mazeh and Shaham (1979), while a higher, third order theory was given by Ford et al. (2000). Briefly, the essence of this effect is as follows: In a hierarchical triple system, where most of the total angular momentum is concentrated in the outer orbit, the eccentricity (e_1), the argument of periastron (g_1) and the inclination (i_1) of the inner orbit are subjects of secular cyclic variations depending upon only one independent variable (one degree of freedom system), namely, the above mentioned argument of periastron (g_1). The mutual inclination of the two orbits is a critical parameter in the equations. If the cosine of the mutual inclination is less than $\sqrt{3/5(1-e_{1_0}^2)}$, where e_{1_0} is the minimal eccentricity of the inner orbit (belonging to $g_1 = 0°$ or $180°$ in case of circulation, or $g_1 = 90°$ or $270°$ in case of libration in g_1, depending upon the initial conditions), then at some values of g_1 its time-derivative becomes zero, and consequently, the time derivative of e_1 becomes constant. Consequently, the eccentricity, e_1 will tend to grow almost linearly in time. (Note, the above value of the critical inclination very weakly depends on other parameters, such as e. g. masses, and the ratio of the separations.) The eccentricity maximum mainly depends on the initial mutual inclination, but very weakly on the initial eccentricity, which means, that in the case of an almost perpendicular initial configuration, the eccentricity of even an originally circular orbit can reach almost 1. After such an eccentricity 'spike', the system returns back to its initial configuration, and a new cycle begins. As it was mentioned, the initial separations and masses hardly affect the eccentricity maximum, only the period of the cycles depends on these quantities. (Note, that according to Ford et al. 2000, including the aforementioned higher order formulae, the cycles are not strictly periodic and similar ones, some bifurcations occur in the $e_1 - \cos g_1$ space, in accordance with the numerical integrations, nevertheless, the chaotic area is usually weak.) Söderhjelm (1984) was the first who showed that other perturbations of the apsidal line (i. e. g_1), like the stellar oblateness forced apsidal motion and/or the relativistic apsidal motion can completely eliminate the Kozai cycles. The possibly important role of Kozai cycles in formation of close binaries first was suggested by Kiseleva et al. (1998). According to this theory the close binaries originally should have had a significantly larger separation, and a distant, inclined third companion. Due to the third star induced Kozai oscillation, the inner eccentricity cyclically becomes so large, that around the periastron passages, the two stars approach each other so closely, that tidal friction may be effective, which then during one ore more Kozai cycles, and some additional circularization after the last cycle, may produce close binaries in their recent forms. Recently Fabrycky and Tremaine (2007) carried out a thorough study on the possibility of the orbital shrinkage by the combined effects of secular perturbations from an inclined distant companion star (Kozai cycles) and tidal friction. They found that „binary stars with orbital periods of 0.1 to 10 days, with a median of ~ 2 d, are produced from binaries with much longer periods (10 d to $\sim 10^5$ d), consistent with observations indicating that most or all short-period binaries have distant companions." What is extraordinarily important in their work is, that they provide some predictions which can be tested by observations. From our point of view the most important one is that in those binaries, where this combined effect could have been effective (namely, for recent periods between 3 and 10 d), the distributions of mutual inclinations of the close and the wide orbits should peak strongly near 40° and 140° (this arises from that feature of Kozai phenomenon, that during the maximum eccentricity phase of the cycle the mutual inclination reaches its minimum which is close to the critical inclination; furthermore, as the tidal friction becomes effective during this period, when the tidal friction cancels the Kozai cycles, it can be expected - and numerical integrations really show this behavior - that this inclination would freeze in the system). Consequently, it is important to determine the mutual inclinations in those triple stars, where the inner period is small enough for the predominancy of tidal forces.

In this context, in the next section we give a short overview of the methods of the determination of the spatial configuration i. e. the mutual inclination of triple stellar systems.

3. Observational aspects of mutual inclination determination

In order to make a distinction between the alternative evolution models on the formation of close binary systems, it is necessary to calculate the relative spatial orientation of the orbits of the close binary and of the tertiary component in as many systems as it is possible. This requires the determination of the spatial orbital elements, namely inclinations (i_1, i_2) and longitude of the nodes (Ω_1, Ω_2) for both the inner and the outer orbits. There are different methods to determine these elements.

(i) The classical one is the astrometric determination. From the sky-projected orbit of a star around its companion all of the six orbital elements can be determined, although the node and the argument of periastron only with an ambiguity of 180°. (Note, that in the case of most interesting small period systems, long-baseline optical and/or radio interferometric observations are required.)

(ii) If the binary is an eclipsing system, then the inclination can be calculated from its light-curve (with the ambiguity, that i, and $180° - i$ give the same solution), but nothing can be said about Ω.

(iii) Finally, a rarely used method comes from polarimetry. If there are some sources of polarized light in the system which revolves together with either of the components, then the phase dependent variation of the polarization could give information both on i and Ω (with the above mentioned ambiguity in this latter).

The ambiguity in the node can be resolved with additional information on the sign of the radial component of the motion (i. e. which component is closer to the observer in a given moment). There are at least three possibilities to get information about the radial component.

(i) The most usual way to get this information arise from radial velocity measurements. Observation of the radial motion of one binary member is already satisfactory for this purpose (single line binaries, SB1). Nevertheless, radial velocity measurements can be used to resolve the node ambiguity only if the two components astrometrically distinguishabe, which is not the case at the simplest long-baseline optical interferometry, where in the absence of phase-information, the two components are interchangeable.

(ii) In the case of an eclipsing binary, the eclipses itself give the most easily observable information about the sign of the (relative) radial coordinate of the two stars. (This is especially true, when the two stars significantly differ which makes perfectly clear that which star is the eclipser and which is the eclipsed.) Nevertheless, the same restriction mentioned above also stands here.

(iii) While the previous two possibilities are mainly applicable for close binaries, in the case of the wide orbit of the third companion, detection of the so-called light-time effect (LITE) could be an easy way to determine the true radial configuration. If the close binary produces any (almost) strictly periodic signal (for example, eclipses, pulsations, ellipsoidal light variations, etc.) while orbiting around the common centre of mass of the triple system, this signal is modulated by the third body, from which the sign of the relative radial coordinate could be easily determined. (Supposing, of course, that the motion has significant radial component.)

Theoretically, the first, astrometric method seems to be the most usable. Unfortunately, the small characteristic size of systems in which we are interested in gives a very strong limit. In the case of a few day-long period binary consisting of two typical stars, the separation of the components is of the order of $0.01 - 0.1$ AU, and consequently, even for a relatively close system with a trigonometric parallax of some $0\farcs01$, the angular size of the orbit is of the order of milliarcseconds (mas). Furthermore, even for the tertiary the typical separation is in the subarcsecond regime. Nevertheless, this latter separation being above the diffraction limit of even smaller, submeter telescopes, by the use of speckle-interferometry or Lucky-imaging, the positions of tertiary companions in several triple systems can be measured with sufficient accuracy for the orbit determination. Turning back to close binaries, the mas, or sub-mas separation gives a very strong practical limit. Such a resolution can be reached only with very long baseline radio interferometry, or the recent, most powerful optical, near-IR long-baseline interferometric equipments, e. g. CHARA array, Palomar Testbed Interferometer and others. Above this direct observational limit, the use of these kind of special instruments gives a further, indirect constraint, which of course, somewhat overlaps with the previous one. Only the brightest close binaries can be reached by such methods. This latter limitation is more expressed in the sense of the radio VLBI measurements narrowing our radio sample only for the nearest chromospherically active binaries. Due to these reasons, the spatial orientation (i. e. the longitude of the node) is known only for very few, approximately a dozen close binaries (see, e. g. Muterspaugh et al. 2006, 2008; Zhao et al. 2008, and further references therein), most of these results were obtained in the last years by optical devices. The first use of VLBI technique in the radio domain for this purpose was carried out for the Algol system by Lestrade et al. (1993). Recently we made a new analysis on the spatial orientation of the Algol system combining the radio (European VLBI Network, EVN) and optical (CHARA array) interferometric measurements (Csizmadia et al. 2009). According to our knowledge, this was the first occasion, when such techniques were applied simultaneously (see next section).

Up to now, we concentrated on the first, astrometric method. Nevertheless, as was mentioned earlier, polarimetric measurements could also reveal information about the spatial orbital elements of close binaries. The method has been known for a long time. It was first described by Shakhovskoi (1965). An improved method was applied for Algol by Rudy and Kemp (1978) and Kemp et al. (1981). Their results are in excellent agreement with the recent CHARA and EVN results (Csizmadia et al. 2009). The great advantage of polarimetry as opposed to long-baseline interferometry is that, it does not require high-tech devices, these measurements could be carried out by significantly smaller, cheaper, and easily reachable instruments. On the other hand, these measurements require some intrinsic, phase-locked polarization in the systems, which condition strongly limits the number of systems measurable by this manner. Although, there are close systems whose spatial orbital elements were determined by this method (for example, some of the most recent measurements are about CQ and CX Cepheis – Villar-Sbaffi et al. 2005, 2006), we are not aware of any systematic surveys to determine spatial orientation of close binaries orbiting in close triple systems by polarimetry.

Determination of the orbital elements of close binaries is only half of the work. The accurate measurements of the wide companion's position is equally important. In most cases this can be done relatively easily even with smaller telescopes by speckle-interferometry. Nevertheless, in most of the cases the orbital period of the third star is of the order of decades (if not centuries). Accordingly, for most of the known systems (listed in the continuously refreshed on-line version of Tokovinin 1997) only a small part of the total orbital arc have been covered, and, consequently, the calculated orbital elements (if they exist) sould be considered only as very preliminary ones. (There are several cases, where elements calculated by different authors within a few years differ completely.) This means that continuous, long-term, accurate observations have essential importance. Furthermore, as it was mentioned, these observations leave a 180° ambiguity in the position of the node, which can be eliminated by the continuous, accurate photometric observations of the eclipsing minima times (supposing that the inner binary is an eclipsing one). In the next section we give a short overview about our new CHARA and EVN measurements about the Algol-system, which was the first step of our intended systematic survey.

4. Interferometric Observations of Algol

4.1. Previuous results

Algol consists of a semi-detached eclipsing binary with an orbital period of 2.87 days (B8V + K2IV) with an F1IV spectral type star revolving around the binary every 680 days. The third component was succesfully observed by speckle interferometry, and its orbit was precisely determined by Bonneau (1979). This result was refined by using the Mark III optical stellar interferometer (Pan et al. 1993).

Considering the spatial orientation of the close binary, the polarimetric measurements of Rudy (1979) yielded $\Omega_1 = 47^\circ \pm 7^\circ$. (Note, that the Ω-ambiguity was also resolved in that work, with the help of the eclipsing phase information.) In the radio regime, Lestrade et al. (1993) detected positional displacement during the orbital revolution of the AB pair using the VLBI technique, and identified the K-subgiant as the source of radio emission. From their measurements Kiseleva et al. (1998) calculated $i_\mathrm{m} = 100^\circ$ for the mutual inclination of the close and wide orbital planes. This value has been widely accepted since then. However, this mutual inclination value cannot be correct due to dynamical considerations, because it would produce a fast variation in the observable inclination of the eclipsing subsystem (see e.g. Borkovits et al. 2004), resulting in a fast eclipse depth variation which contradicts to the more than century-long photometric observations (Söderhjelm 1980).

4.2. Observations and results

We carried out interferometric observations both in the optical/near-infrared, and in the radio regime. The optical interferometry observations were done by the CHARA Array on three nights (2, 3 and 4 December, 2006) in the K_s band. The radio measurements were carried out with a subset of the European VLBI Network (EVN) on 14-15 December 2006 at 5 GHz. The participating telescopes were Cambridge and Jodrell Bank (UK), Medicina (Italy), Onsala (Sweden), Toruń (Poland) and the Westerbork phased array (the Netherlands). The measurements and the data reductions were carried out according to the standard procedures (for details see Csizmadia et al. 2009).

From the analysis of the CHARA visibilities we determined the angular size of the binary's semi-major axis (a_1), the surface brightness ratio in K_s band (J_{K_s}), and the angle Ω_1, as follows $a_1 = 2.28 \pm 0.02$mas, $J_{K_s} = 0.330 \pm 0.01$, and $\Omega_1 = (48^\circ + k \cdot 180^\circ) \pm 2^\circ$ $(k = 0, 1)$, respectively.

The VLBI data were gained during a secondary minimum, which, according to our simultaneous optical photometry, occurred at $t_0 = 2\,454\,084.360 \pm 0.003$ (Bíró et al. 2007). The measurements lasted 9 hours, i.e. $\approx 13\%$ of the orbital period, but as the projected movement of the source is maximal around the minima, the covered projected orbital arc is significantly larger. Due to the large inclination, this projected arc is almost a straight line parallel to the nodal line, which makes the determination of the longitude of the node easier. Our best fit gave $\Omega_1 = 52^o \pm 3^o$, resolving the node-ambiguity of the CHARA observations. However, the Ω value itself is likely less reliable, because the source flared during our observations, in which case structural variations (not reflecting orbital motion) are likely to happen. (Note, that the displacement of the source during our observation as almost twice of what was expected from Keplerian revolution.)

Due to the above mentioned facts, here we concentrate mainly on the CHARA results. The true size of the obtained semi-major axis and surface brightness ratio are in very good agreement with previous results, which makes our nodal result also plausible. The obtained $\Omega_1 = 48^o \pm 2^o$ is in excellent agreement with the value determined from polarimetric measurements mentioned above, indicating that polarimetry is an efficient tool to determine the spatial orientation of the orbits.

At this point, using the well-known cosine theorem of spherical triangles

$$\cos i_{\rm m} = \cos i_1 \cos i_2 + \sin i_1 \sin i_2 \cos(\Omega_1 - \Omega_2) \qquad (1)$$

we can determine the mutual inclination $i_{\rm m}$, and we get $i_{\rm m} = 95^o \pm 3^o$. This value is, however, closer to the exact perpendicularity than the 100° which was based on the measurements of Lestrade et al. (1993). Note, that for the first sight, the difference by 5° does not seem to be crucial, nevertheless, because the period of the precession of the orbital plane is approximately proportional to $\cos^{-1} i_{\rm m}$ (see e. g. Söderhjelm, 1975; or for both the dynamical and the observable node in Borkovits et al. 2007), consequently, this new result gives a longer period by a factor of two which is already consistent with the observations.

5. Conclusions

The recent spatial configuration of close binary systems might be accurate tracers of the evolution and formation of such systems in the past, and indirectly of the stellar formation theories. Nowadays we already have the possibility to see inside of such systems. In this paper we gave a short overview about some of the practical, technical, strategic questions of aquiring such results, as well as some aspects of the theoretical predictions behind them. Considering the actually studied system, Algol, we found that results yielded with different technics can support each other. Nevertheless, we note, that the triple system of Algol itself is not the most ideal subject for the investigation of the consequences of the Kozai mechanism. This is the case at least for two reasons. (i) The secondary component of Algol is an evolved star, consequently, the system had passed on the Case A mass-transfer phase in the past which should have changed the configuration, including the separation of the binary dramatically. (ii) Furthermore, the third companion is too close to the binary (at least from this specific point of view) with respect to the typical triple stellar systems, and consequently, the general statistical considerations applied in Fabrycky and Tremaine (2007) might be not perfectly valid.

In the following we plan a systematic survey of hierarchical triple stellar systems to determine the mutual spatial configurations, in order to be able to make a statistical investigation of the relative orbits.

References

Batten A H 1973 *Binary and multiple systems of stars* (Oxford: Pergamon Press)
Bíró I B, Borkovits T, Hegedüs T et al. 2007 *IBVS* **5753** 1
Bonneau D 1979 *A&A* **80** 11
Borkovits T, Forgács-Dajka E, Regály Zs 2004 *A&A* **426** 951
Borkovits T, Forgács-Dajka E, Regály Zs 2007 *A&A* **473** 191
Csizmadia Sz, Borkovits T, Paragi Zs, Ábrahám P, Szabados L, Mosoni L, Sturmann L, Sturmann J, Farrington C, McAlister H A, ten Brummelaar T A, Turner N H and Klagyivik P 2009 *ApJ* **705** 436

Journal of Physics: Conference Series **218** (2010) 012005 doi:10.1088/1742-6596/218/1/012005

D'Antona F and Mazzitelli I 1994 *ApJS* **90** 467
Eggleton P P and Tokovinin A A 2008 *MNRAS* **389** 869
Fabrycky D and Tremaine S 2007 *ApJ* **669** 1298
Ford E B, Kozinsky B and Rasio F A 2000 *ApJ* **535** 385
Kemp J C, Barbour M S, McBirney R E and Rudy R J 1981 *ApJ* **243** 557
Kiseleva L G, Eggleton P P and Mikkola S 1998 *MNRAS* **300** 292
Kozai Y 1962 *AJ* **67** 591
Lestrade J-F, Phillips R B, Hodges M W and Preston R A 1993 *ApJ* **410** 808
Mazeh T and Shaham J 1979 *A&A* **77** 145
Muterspaugh M W, Lane B F, Konacki M, Burke B F, Colavita M M, Kulkarni S R and Shao M 2006 *A&A* **446** 723
Muterspaugh M W, Lane B F, Fekel F C, Konacki M, Burke B F, Kulkarni S R, Colavita M M, Shao M and Wiktorowicz S J 2008 *AJ* **135** 766
Pan X, Shao M and Colavita M M 1993 *ApJ* **413** 129
Rudy R J 1979 *MNRAS* **186** 473
Rudy R J and Kemp J C 1978 *ApJ* **221** 200
Shakhovskoi N M 1965 *Sov. Astr.* **8** 833
Söderhjelm S 1975 *A&A* **42** 229
Söderhjelm S 1980 *A&A* **89** 100
Söderhjelm S 1984 *A&A* **141** 232
Tokovinin A A 1997 *A&AS* **124** 75
`http://www.ctio.noao.edu/~atokovinin/stars/index.php`
Tokovinin A A, Thomas S, Sterzik M and Udry S 2006 *A&A* **450** 681
Villar-Sbaffi A, St-Louis N, Moffat A F J and Piirola V 2005 *ApJ* **623** 1092
Villar-Sbaffi A, St-Louis N, Moffat A F J and Piirola V 2006 *ApJ* **640** 995
Zhao M, Gies D, Monnier J D, Thureau N, Pedretti E, Baron F, Merand A, ten Brummelaar T, McAlister H, Ridgway S T et al. 2008 *ApJ* **684** L95

Journal of Physics: Conference Series **218** (2010) 012006 doi:10.1088/1742-6596/218/1/012006

Numerical investigations of gravitational collapse*

Péter Csizmadia, István Rácz

RMKI, Budapest, Konkoly Thege Miklós út 29-33, H-1121 Hungary

E-mail: iracz@rmki.kfki.hu

Abstract. Some properties of a new framework for simulating generic 4-dimensional spherically symmetric gravitating systems are discussed. The framework can be used to investigate spacetimes that undergo complete gravitational collapse. The analytic setup is chosen to ensure that our numerical method is capable to follow the time evolution everywhere, including the black hole region.

1. Introduction

All those people who has far reaching memory about the history of ideas in general relativity may recall the time when cosmic censor hypotheses were formulated by Penrose. The strong cosmic censor hypothesis expresses our expectation that generic spacetimes are globally hyperbolic (Penrose 1979). This, in particular, rules out the formation of Cauchy horizons and, in turn, it is expected to justify that there is no limitation on the predictive power of general relativity. One of the famous drawings by Penrose (see Fig. 1) depicts the diagram of a cosmological situation where spacetime is expanding steadily but black holes are also formed. According to results of

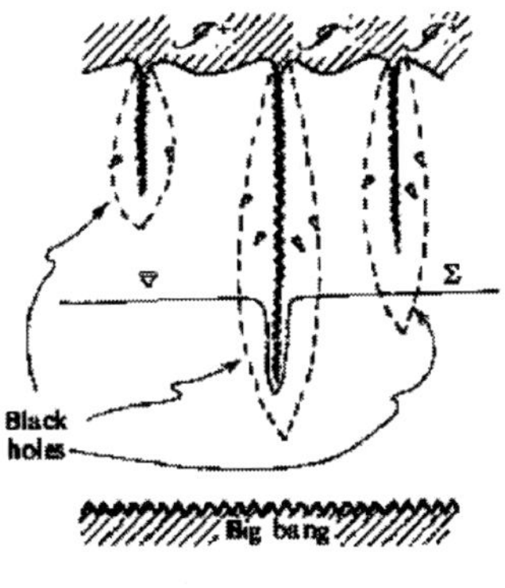

Geroch (Geroch 1967, 1970) the base manifold M of such a globally hyperbolic spacetime—i.e., the Cauchy development of some three-dimensional achronal hypersurface Σ—has to possess the product space structure $\Sigma \times \mathbb{R}$. Then the entire spacetime may be foliated by suitably chosen Cauchy surfaces that are diffeomorphic to each other. Whenever black holes are formed the Cauchy slices should bend around them—as it is suggested by the drawing of Penrose—in order to get them close to and simultaneously keep them away from the developing singularity.

* Dedicated to the memory of Péter Csizmadia who was missing–along with three other young Hungarian climbers–at the end of October 2009 while taking part in the first ascent attempt of Ren Zhong Feng in China.

It is worth to recall now that in numerical relativity there is a huge number of simulations where black holes play significant role. However, in most of these numerical investigations, one of the aims was to avoid entering to the trapped regions. This was achieved by making use of techniques of trapped region avoiding slicing and black hole excision (Thornburg 1987, Seidel 1992). The unsatisfactory aspect of this trouble avoiding attitude is that—opposed to the strong cosmic censor hypothesis—we cannot investigate evolution everywhere. This limitation is what we intended to overcome and what was successfully done for the first time in (Csizmadia 2010). In doing so we applied two tricks. The first one provided us the chance to have stable evolution in the trapped region while the second one was applied to get the largest possible part of the spacetime be covered by the time slices. Although the concept of "maximal Cauchy development" was introduced in (Choquet-Bruhat 1969). However, in proving the existence of the largest possible Cauchy development Zorn's Lemma has to be used. Therefore, it is of crucial importance to have a practical method providing us the largest possible Cauchy development.

2. The dynamical system

For simplicity we consider only the evolution of a real self-interacting scalar field with Lagrangian $\mathscr{L} = \frac{1}{2}g^{ef}\nabla_e\psi\nabla_f\psi - V$ in Einstein's theory of gravity. Accordingly the energy-momentum tensor is $T_{ab} = \nabla_a\psi\nabla_b\psi - g_{ab}\left[\frac{1}{2}g^{ef}\nabla_e\psi\nabla_f\psi - V\right]$ and the field equations are

$$E_{ab} = R_{ab} - \frac{1}{2}g_{ab}R - 8\pi T_{ab} = 0\,, \tag{1}$$

$$\nabla_e\nabla^e\psi + \frac{\partial V}{\partial\psi} = 0\,. \tag{2}$$

The metric of the 4-dimensional spherically symmetric spacetime (M, g_{ab}) was chosen to be

$$ds^2 \;=\; \alpha\,\beta^2\,d\tau^2 - \alpha\,d\rho^2 - r^2\left(d\theta^2 + \sin^2\vartheta\,d\varphi^2\right)\,, \tag{3}$$

where the coordinates τ and ρ label the points of the two-dimensional timelike surfaces transverse to the transitivity surfaces of the rotation group, and α, β, r are smooth functions of $(\tau,\,\rho)$.

Before providing the basic equations satisfied by the dynamical variables α, r and ψ let us recall that in spherically symmetric spacetimes any $(1,1)$-type tensor field $E^\mu{}_\nu$ must have the structure

$$E^\mu{}_\nu = \begin{pmatrix} E^\tau{}_\tau & E^\tau{}_\rho & 0 & 0 \\ E^\rho{}_\tau & E^\rho{}_\rho & 0 & 0 \\ 0 & 0 & E^\vartheta{}_\vartheta & 0 \\ 0 & 0 & 0 & E^\varphi{}_\varphi \end{pmatrix}, \tag{4}$$

where $E^\varphi{}_\varphi = E^\vartheta{}_\vartheta$. If in addition $E_{\mu\nu}$ is symmetric and the metric is given by the line element (3) then we also have $E^\rho{}_\tau = -\beta^2 E^\tau{}_\rho$. Whenever E_{ab} is defined by the r.h.s. of (1), the evolution equations for α, r can be seen to be given by the combinations

$$E^\varphi{}_\varphi + E^\vartheta{}_\vartheta - E^\rho{}_\rho - E^\tau{}_\tau = 0\,, \qquad E^\rho{}_\rho + E^\tau{}_\tau = 0\,, \tag{5}$$

while the constraint equations read as

$$E^\tau{}_\rho \;=\; 0, \qquad E^\tau{}_\tau \;=\; 0\,. \tag{6}$$

Thus (5) and (6) are equivalent to Einstein's equations, i.e., requiring the four non-identically zero components of $E^\mu{}_\nu$ to vanish. It can also be shown that, if the evolution equations hold,

$$\partial_\tau E^\tau{}_\rho - \partial_\rho E^\tau{}_\tau + \frac{2}{r}\left[r_\tau E^\tau{}_\rho - r_\rho E^\tau{}_\tau\right] + \frac{(\alpha_\tau\beta + \alpha\beta_\tau)E^\tau{}_\rho - (2\alpha_\rho\beta + \alpha\beta_\rho)E^\tau{}_\tau}{\alpha\beta} \;=\; 0, \tag{7}$$

$$\partial_\tau E^\tau{}_\tau - \beta^2\partial_\rho E^\tau{}_\rho + \frac{2}{r}\left[r_\tau E^\tau{}_\tau - \beta^2 r_\rho E^\tau{}_\rho\right] + \frac{\alpha_\tau E^\tau{}_\tau - \beta(\alpha_\rho\beta + 3\alpha\beta_\rho)E^\tau{}_\rho}{\alpha} \;=\; 0 \tag{8}$$

are satisfied. Since these equations are linear and homogeneous in the variables $E^{\tau}{}_{\tau}$ and $E^{\tau}{}_{\rho}$ they have identically zero solution for vanishing initial data, i.e., the constraints propagate whence it is completely satisfactory to impose them only on the initial data hypersurface.

2.1. *The field equations*

The simpleminded use of the above setup leaded to unstable evolutions because close to an origin the expression $\frac{\alpha\beta^2+r_\tau^2-\beta^2 r_\rho^2}{r^n}$—that appears in (5), with $n=1$ and $n=2$, respectively—are of $0/0$ type. To cure these instabilities it was crucial to use the Misner-Sharp mass m

$$m = \frac{r}{2}\left(1+g^{ab}\partial_a r\,\partial_b r\right) = \frac{r}{2}\left(\frac{\alpha\beta^2+r_\tau^2-\beta^2 r_\rho^2}{\alpha\beta^2}\right). \tag{9}$$

By applying m as an additional new variable the basic field equations could be given as

$$\partial_\tau\alpha_\tau = \beta^2\partial_\rho\alpha_\rho + \frac{4m\beta^2\alpha^2}{r^3} + 8\pi\alpha^2\beta^2\left(T^{\vartheta}{}_{\vartheta}+T^{\varphi}{}_{\varphi}-T^{\tau}{}_{\tau}-T^{\rho}{}_{\rho}\right) + \frac{\alpha_\tau^2-\beta^2\alpha_\rho^2}{\alpha} + \frac{\beta_\tau\alpha_\tau+\beta^2\beta_\rho\alpha_\rho}{\beta} + 2\alpha\beta\partial_\rho\beta_\rho\,, \tag{10}$$

$$\partial_\tau r_\tau = \beta^2\partial_\rho r_\rho + 4\pi r\alpha\beta^2\left(T^{\tau}{}_{\tau}+T^{\rho}{}_{\rho}\right) - \frac{2m\beta^2\alpha}{r^2} + \frac{\beta_\tau r_\tau+\beta^2\beta_\rho r_\rho}{\beta}, \tag{11}$$

$$\partial_\tau m = 4\pi r^2\left(r_\tau T^{\rho}{}_{\rho} - r_\rho T^{\rho}{}_{\tau}\right), \tag{12}$$

$$\partial_\tau\psi_\tau = \beta^2\partial_\rho\psi_\rho - 2\frac{\psi_\tau r_\tau-\beta^2\psi_\rho r_\rho}{r} + \frac{\beta_\tau\psi_\tau+\beta^2\beta_\rho\psi_\rho}{\beta} - \alpha\beta^2\frac{\partial V}{\partial\psi}, \tag{13}$$

where the abbreviating notation $f_\sigma = \partial_\sigma f$ is applied (for $f=\alpha$, β, r and ψ). Notice that there is no evolution equation for β.

The full set of evolution equations can be shown (Csizmadia 2010) to possess the form of a first order strongly hyperbolic system $\partial_\tau\mathbf{u} = \mathbf{A}\partial_\rho\mathbf{u}+\mathbf{B}$ for the 10-dimensional vector variable $\mathbf{u} = (m,\alpha,\alpha_\tau,\alpha_\rho,r,r_\tau,r_\rho,\psi,\psi_\tau,\psi_\rho)^T$ and with a suitable source term $\mathbf{B}$, the nontrivial components of which can be determined by making use of (10), (11), (12) and (13). Note that to these type of evolution equations the existence and uniqueness of solutions is guaranteed.

3. Numerical results

The first order strongly hyperbolic system of evolution equations were solved numerically with our finite difference code called *GridRipper AMR*, within which the time integration is done by making use of the method of lines based on a fourth order Runge-Kutta scheme. *GridRipper* does also incorporate the techniques of adaptive mesh refinement [Csizmadia 2007].

3.1. *The tricks*

Before presenting the our first trick that enabled us to enter to the trapped region let us recall that—as it was shown in (Csizmadia 2010)—a 2-sphere $\mathscr{S}$ of radius r, that is invariant under the action of the $SO(3)$ symmetry group, is untrapped or trapped (future or past) if and only if $1-\frac{2m}{r}$ is positive or negative, respectively. This characterization follows from the fact that the principal null expansions $\theta_\pm$—with respect to the principal null directions $n^a_\pm = \left(\frac{\partial}{\partial\tau}\right)^a \pm \beta\left(\frac{\partial}{\partial\rho}\right)^a$—can be related to the Misner-Sharp mass as

$$1-\frac{2m}{r} = \frac{\beta^2 r_\rho^2 - r_\tau^2}{\alpha\beta^2} = -\frac{r^2}{4\,\alpha\beta^2}\cdot\theta_+\theta_-\,. \tag{14}$$

Our first "trick" was nothing more than keeping the evolution equation (10) for α instead of determining it by the relation $\alpha = \left(1-\frac{2m}{r}\right)^{-1}\frac{\beta^2 r_\rho^2 - r_\tau^2}{\beta^2}$ as in the latter case we had fast growing numerical errors at the marginally trapped surfaces.

The second trick enabling us to get the largest possible Cauchy development was the introduction of a dynamical lapse function. As β is freely specifiable we chose it to evolve according to the equation

$$\beta_\tau = -p\left[\frac{m_\tau r - 3\,m\,r_\tau}{r^4}\right]\left(\frac{r_\tau}{r}\right)^3\beta\,, \tag{15}$$

where p is a positive real parameter, starting with the initial value $\beta|_{\Sigma_0} = 1$. This way we were able to slow down the evolution at those parts of the time level surfaces which got too close to the singularity providing thereby apparently the largest possible spacetime domain associated with the chosen initial time slices.

3.2. The dynamics of time slices

In representing the yielded spacetime the use of the conformal time $\tau_c = \int_0^\tau \beta d\tau$ turned to be advantageous since the radial null geodesics are well approximated by $\tau_c \pm \rho = const$ lines. Its definition comes from the fact that along radial null geodesics $\beta d\tau \pm d\rho = 0$ holds. This, along with $d\tau_c = \beta d\tau + \left[\int_0^\tau \beta_\rho d\tau\right] d\rho$ and that the τ-integral of β_ρ is negligible, implies that in the $\rho - \tau_c$ coordinate plane for the radial null geodesics $d\tau_c/d\rho = \mp 1 + \left[\int_0^\tau \beta_\rho d\tau\right] \approx \mp 1$ holds.

The spacetime diagrams below depict the time evolution relevant for initial data surfaces with topology $\Sigma_0 = \mathbb{R}^3, \mathbb{S}^3$ and $\mathbb{S}^1 \times \mathbb{S}^2$. The applied initial data in each of these cases is exactly the same as it was in the corresponding simulations of (Csizmadia P 2010). The main point here is the way the $\tau = const$ time level surfaces—indicated by the thin dotted lines—bend as the evolution slows down close to the $r = 0$ singularity—represented by the thick dashed lines—while the evolution remains regular everywhere else. The future and past apparent horizons are also indicated by the thick and thin continuous lines.

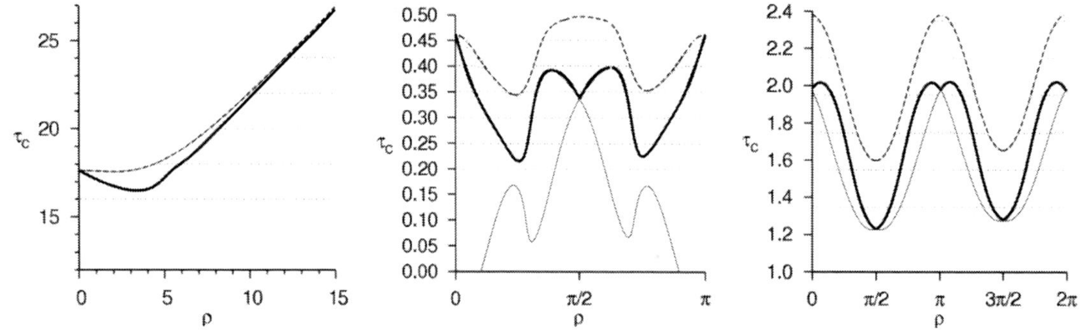

Acknowledgments

This research was supported in part by OTKA grant K67942.

References

Penrose R 1979 *General Relativity: An Einstein Centenary Survey*, eds. Hawking and Israel
Geroch R 1967 *J. Math. Phys.* **8** 782-6
Geroch R 1970 *J. Math. Phys.* **11** 437-49
Thornburg J 1987 *Class. Quantum Grav.* **4** 1119-31
Seidel E and Suen W-M 1992 *Phys. Rev. Lett.* **69** 1845-9
Csizmadia P and Rácz I 2010 *Class. Quantum Grav.* **27** 015001
Choquet-Bruhat Y and Geroch R 1969 *Commun. Math. Phys.* **14** 329-35
Csizmadia P 2007 *Class. Quantum Grav.* **24** S369

5th Workshop of Young Researchers in Astronomy & Astrophysics IOP Publishing
Journal of Physics: Conference Series **218** (2010) 012007 doi:10.1088/1742-6596/218/1/012007

Linear waves on fixed Kerr background and their relevance in jet formation *

Péter Csizmadia[1], András László[1,2], István Rácz[1]

[1]RMKI, Budapest, Konkoly Thege Miklós út 29-33, H-1121 Hungary
[2]CERN, Geneva, Switzerland.

E-mail: laszloa@rmki.kfki.hu, iracz@rmki.kfki.hu

Abstract. In the conventional models of astrophysical jets magnetohydrodynamics is believed to play the most role. Recently there were several qualitative arguments emphasizing that the role of pure gravitational effects might be more important than expected before. Here we present some preliminary but quantitate results clarifying the role of gravity in the formation of astrophysical jets.

1. Introduction

A great number of astrophysical observations, see, e.g, [Mirabel&Rodriguez1994], [Tingay1995], justified the existence of high energy highly collimated matter streams. The sources are usually compact celestial objects while the yielded matter currents are referred to as astrophysical jets. The conventional explanation of their existence is based on the use of magnetohydrodynamics. The emitting compact object is assumed to possess strong magnetic field which interacts with the surrounding ionized matter. The corresponding investigations justified the formation of matter currents confined to the axis of the magnetic field (see, e.g., Marti 2003).

Recently, there were several authors proposing alternative explanations for the formation of jets where gravitational effects were assumed to play central role. For instance, in Gariel (2007) the following possibility, making use of the Penrose process, was considered. A steady fall of particles of an accretion disc towards the central Kerr black hole was considered and the Penrose process (see below) was used to explain the formation of strongly collimated streams of outgoing particles along the axis of rotation. However, in Gariel (2007) beside introducing a possible mechanism no quantitative investigations were made whence no reliable estimation on the relevance of the proposed effect could be provided. The main motivation for the present authors was to derive these type of estimates by studying the evolution of a linear field on a fixed Kerr background.

2. Penrose process and super-radiance in Kerr black hole spacetime

It is well-known (see, e.g., Wald 1984) that each rotating Kerr black hole possesses a subregion – called the ergoregion of the black hole – in the domain of outer communication, where the

* This article is dedicated to the memory of our friend and colleague Péter Csizmadia, the original author of the GridRipper. Péter Csizmadia disappeared in October 2009 during a first ascent attempt of the unclimbed peak Ren Zhong Feng in the Chinese Himalayas.

asymptotically timelike Killing vector field t^a is spacelike. In this region, the energy $g_{ab}t^a p^b$ of a point particle with timelike momentum vector p^a may become negative. The Penrose process can be used to extract energy from a rotating Kerr black hole as follows. Shoot an unstable point particle into the ergoregion which is prepared such that it decays into two particles there with one of them having negative energy. This negative energy particle cannot exit the ergoregion as the energy of point particle in the asymptotic region has to be positive. Thereby, it is captured by the black hole. Then, in virtue of the energy conservation, the other yielded particle, has to have positive energy – in fact larger that that of the original particle sent into the ergoregion – whence it escapes to infinity. What was shown in Gariel (2007) is that in a rotating Kerr black hole spacetime there exists a class of causal geodesics, the elements of which start at the ergoregion, and leaves the ergoregion asymptotically parallel to the rotation axis. If these trajectories were used by the escaping particles of the Penrose process one could get a viable explanation of jets.

Recall now that the super-radiance provides a much less ambiguous process of energy extraction from black hole spacetimes. As shown in Carter (1968), the d'Alembert operator over Kerr spacetime may be separated in Boyer-Lindquist coordinates t, r, ϑ, φ. The timelike Fourier transform of a solution may be written in the form

$$\sum_{l=0}^{\infty} \sum_{m=-l}^{l} R^m_{l,\omega}(r) S^m_{l,\omega}(\vartheta, \varphi),$$

where ω is the frequency in the time translation direction, $S^m_{l,\omega}$ is the oblate spheroidal harmonic function with oblateness parameter ω and with angular momentum quantum numbers l and m. The $R^m_{l,\omega}$ radial functions may be determined by the radial field equations. If one takes a plane wave solution of the form $\exp(-\mathrm{i}\omega t)\phi(r, \vartheta)\exp(\mathrm{i}m\varphi)$, then the energy for a massless Klein-Gordon field absorbed through the event horizon can be written as

$$\frac{1}{2}\omega(\omega - m\Omega_H)|\phi|^2,$$

where Ω_H is the angular velocity of the black hole (Wald 1984). Thus the energy inflow is negative, whenever $0 < \omega < m\Omega_H$. If this happens the backscattered wave carries more energy than the wave sent towards the black hole. This field theoretical analogue of the Penrose process is called super-radiance which is known to occur for all matter fields satisfying linear wave equations.

3. The anisotropy of linear waves on Kerr background

This paper is to clear up the significance of collimation effect proposed by Gariel (Gariel 2007). In doing so we shall consider the time evolution of massless Klein-Gordon field on fixed rotating black hole spacetime where super-radiance may occur. The applied numerical method provides the opportunity of making quantitative investigations.

3.1. The numerical method

To carry out our investigations, we developed the 3+1 dimensional version of the PDE solver, GridRipper (Csizmadia 2009, Csizmadia 2006, Csizmadia 2007). The former version of GridRipper was an 1+1 dimensional fourth order finite difference PDE solver with adaptive mesh refinement (AMR) for solving hyperbolic equations. It is based on the Berger-Oliger algorithm: refinement is performed whenever the Richardson error of a predefined field quantity exceeds a threshold. In the 3+1 dimensional version the polar directions are taken into account by expanding the fields with respect to the L^2 basis of spherical harmonics, and the spectral components up to $l = 24$ are evolved. Since on a rotating Kerr spacetime the wave equation immediately involves multilinear expressions of the multipole components the applicability of involved spectral method is not immediately obvious. The applicability of spectral methods may be justified by applying Sobolev embedding arguments which can be found in Csizmadia (2010).

3.2. Results and discussion

As indicated above we considered the evolution of the massless Klein-Gordon equation on a fixed Kerr spacetime. The black hole is parameterized by its mass M and its specific angular momentum a. The initial data was assumed to be radially non-expanding possessing the form

$$\Phi|_{t=0} = \exp(-i\omega t)\,\phi(r,\vartheta)\exp(im\varphi)|_{t=0}\,,$$

$$\partial_t\Phi|_{t=0} = \partial_t\left(\exp(-i\omega t)\,\phi(r,\vartheta)\exp(im\varphi)\right)|_{t=0}\,,$$

where the profile of ϕ was chosen such that the initial condition represented a rotating toroidal ring. The parameter ω was non-negative real, while m was chosen to be an integer. The energy density, angular momentum density and energy current density distributions are shown on Fig. 1, relevant for a configuration with parameters $M = 1$, $a = 0.5$, $\omega = 1$, $m = 1$. On the right panel an apparent preference of the axis of rotation is visible.

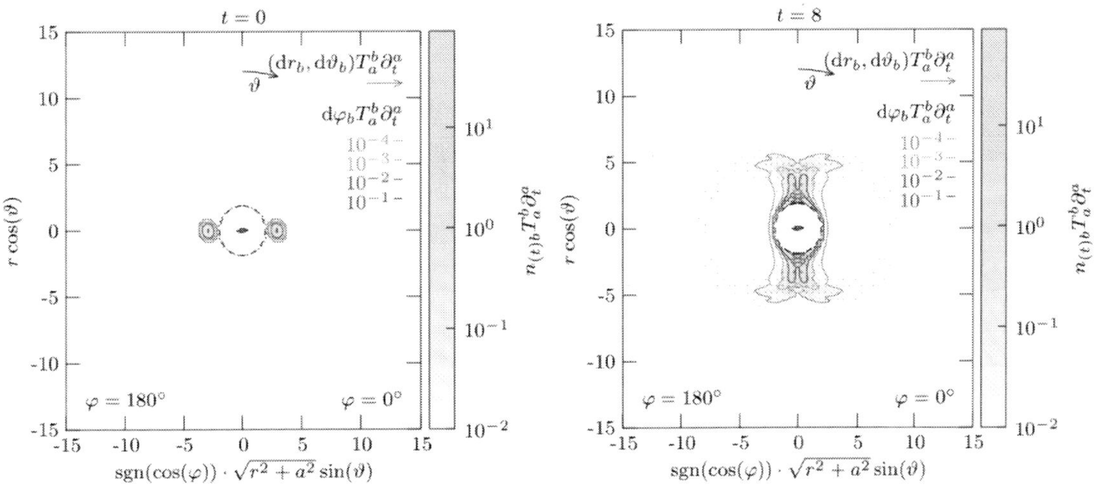

Figure 1. Initial condition at $t = 0$ (left panel) and intermediate state at $t = 8$ (right panel) of a massless Klein-Gordon field over Kerr background spacetime, with parameters $M = 1$, $a = 0.5$, $\omega = 1$ and $m = 1$.

To characterize the radiation anisotropy, the following measure of anisotropy was used. Denote by $E_{\mathrm{out}}|_{\overline{\vartheta}}(t)$ the outwards radiated energy within the cone with half-angle $\overline{\vartheta}$, $0 \le \overline{\vartheta} \le \pi$, measured at an outer boundary as the function of coordinate time t. Then the quantity

$$\mathcal{E}_{\overline{\vartheta}}(t) = \frac{4\pi}{2\pi(1-\cos(\overline{\vartheta}))}\cdot\frac{E_{\mathrm{out}}|_{\overline{\vartheta}}(t)}{E_{\mathrm{out}}|_{\pi}(\infty)}$$

provides a measure of angular preference of the energy of the radiation. $\mathcal{E}_{\overline{\vartheta}}(t)$ tends to 0 whenever the radiation prefers the equatorial directions, whereas it tends to 1 in the case of spherically symmetric radiation, while $\mathcal{E}_{\overline{\vartheta}}(t)$ tends to a value larger than 1 when the radiation prefers the axis of rotation. Obviously, an analogous quantity $\mathcal{L}_{\overline{\vartheta}}(t)$ can also be constructed to measure the angular momentum anisotropy of the radiation.

The preliminary results of our numerical investigations are shown on Fig. 2 for the specific choice $\overline{\vartheta} = 30°$ and for spacetimes $M = 0$, $a = 0$ (Minkowski limit), $M = 1$, $a = 0$ (Schwarzschild limit) and $M = 1$, $a = 0.5$ (a rotating Kerr spacetime). The parameters of the initial field configurations are $\omega = 1$, $m = \pm 1$ (for co-rotating or counter rotating field) and $\omega = 1$, $m = 0$ (for non-rotating field). As it is shown by these panels the energy radiation anisotropy is a little bit higher for the case of black hole spacetimes. However, as the energy anisotropy measured in case of Schwarzschild or Kerr black holes are almost the same it cannot be attributed to the super-radiance. The slight difference indicated by the right bottom panel justifies that the rotation of the background yields an anticipated asymmetry in the angular momentum anisotropy. These results do not seem to support the method proposed in Gariel (2007).

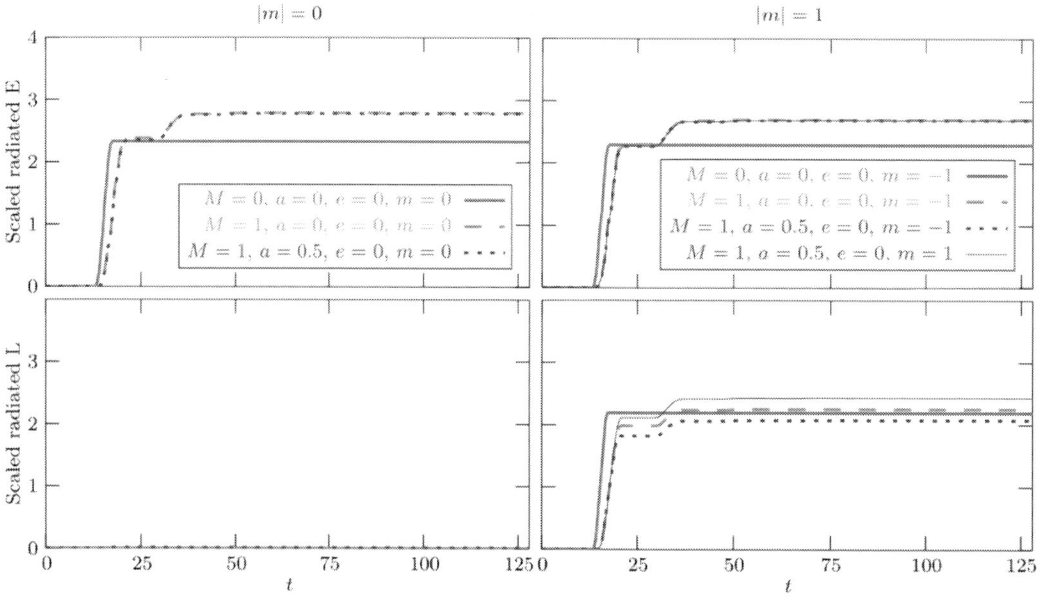

Figure 2. Energy (top panels) and angular momentum (bottom panels) radiation anisotropy in the $\overline{\vartheta} = 30°$ cone, for initially non-rotating (left panels) and initially rotating (right panels) fields are shown.

Acknowledgments

This research was supported in part by OTKA grant K67942.

References

Carter B 1968 *Phys. Rev.* **174** 1559
Csizmadia P, László A and Rácz I 2009 The GridRipper homepage (`http://www.kfki.hu/~cspeter/gridripper`)
Csizmadia P 2006 *Int. J. Mod. Phys.* **D15** 107
Csizmadia P 2007 *Class. Quant. Grav.* **24** S369
Csizmadia P, László A and Rácz I 2010 in preparation
Gariel J, MacCallum M A H, Marcilhacy G and Santos N O 2007 Kerr geodesics, the Penrose process and jet collimation by a black hole (Preprint gr-qc/0702123)
Marti J M and Müller E 2003 Numerical hydrodynamics in special relativity *Living Reviews in Relativity* **6**
Mirabel I F and Rodriguez L F 1994 *Nature* **371** 4648
Tingay J S et al. 1995 *Nature* **374** 141143
Wald R M 1984 *General Relativity* (Chicago: University of Chicago Press)

5th Workshop of Young Researchers in Astronomy & Astrophysics IOP Publishing
Journal of Physics: Conference Series **218** (2010) 012008 doi:10.1088/1742-6596/218/1/012008

Some preliminary results on the dynamics of Einstein–Yang–Mills–Higgs systems *

Péter Csizmadia[2], **Gergely Kovács**[1,2], **István Rácz**[2]

[1] Eötvös University, Department of Astronomy, Budapest, POB. 32, H-1518 Hungary
[2] RMKI, Budapest, Konkoly Thege Miklós út 29-33, H-1121 Hungary

E-mail: gkovacs@rmki.kfki.hu, iracz@rmki.kfki.hu

Abstract. A mathematical and numerical framework for the study of dynamical properties of spherically symmetric Einstein–Yang-Mills–Higgs systems is introduced. The quantitative investigations of the time evolution of some simple magnetic monopole type configurations are presented. Long living breather type states are found to develop.

1. Introduction

The study of the evolution of nonlinear dynamical systems, in particular, that of the properties of radiative tails, is at the center of various analytic and numerical investigations (see, e.g., references Zenginoglu 2008, Pürrer & Aichelburg 2009, Bizoń et al. 2009, Bizoń & Zenginoglu 2009). Although most of these studies are restricted to spherically symmetric configurations—due to the fact that some of the involved fields are radiative—the results beside being of interest on their own rights have also some relevance concerning the plausible fall off properties of gravitational waves. In order to get the gravitational degrees of freedom to become dynamical one needs to use matter fields whenever the geometry and the fields are spherically symmetric. The main motivation for the study of these type of systems is provided by the fact that the linearization of a nonlinear system fails to predict the correct asymptotic behavior of the involved fields—see, e.g., Table 1 below—even in regions far away from the truly dynamical and visible nonlinear regions.

In this paper the dynamics of Einstein–Yang-Mills–Higgs systems is investigated with the assumption that the Yang-Mills field is in the adjoint representation of the $SU(2)$ gauge group. Our eternal aim will be the study of the asymptotic properties of these systems. Nevertheless in this paper considerations will be restricted to the investigation of dynamics in the inner region containing a localized particle type soliton.

The Einstein–Yang-Mills–Higgs systems—beside being of great interest from the point of view of nonlinear dynamics—are known to host various particle like solutions discovered by Bartnik & McKinnon 1988 and later these static systems were investigated by many authors see, e.g., Bizoń 1990, Breitenlohner et al. 1994. The reader interested in these more particle physics oriented motivations may find the review papers Goddard & Di Olive 1978, and Volkov & Gal'tsov 1999 to be useful.

* Dedicated to the memory of our co-author and friend Péter Csizmadia who was missing–along with three other young Hungarian climbers–during an expedition in Himalayas at the end of October 2009.

Table 1. The value of the power p of the $\sim A\tau^p$ type fall off law of the matter field variable is indicated while approaching future timelike infinity i^+ and future null infinity $\mathscr{I}^+$, respectively, where in both cases τ denotes the coordinate time of the pertinent spacetime. In all of these investigations the evolution of a single Yang-Mills field was considered.

i^+	$\mathscr{I}^+$	Spacetime	Technique
-5	-3	Schwarzschild	Linear Perturbation
-4	—	Schwarzschild and Minkowski	Nonlinear Perturbation
-4	-2	Schwarzschild	Numerical Method
-4	-2	Spherically Symmetric	Numerical Method

2. The gauge choices

This section is to provide a precise specification of the gravity matter model to which our results apply.

2.1. *The spacetime*

Regarding the geometry we shall adopt the choices made in Csizmadia & Rácz 2010. Accordingly the metric of the spherically symmetric spacetime is supposed to be given by line element

$$ds^2 = \alpha\beta^2 d\tau^2 - \alpha d\rho^2 - r^2 dS^2 \tag{1}$$

where α, β and r are smooth functions of the coordinates τ, ρ, and dS^2 stands for the line element of the two-dimensional unite sphere, i.e., $dS^2 = d\vartheta^2 + \sin^2\vartheta\, d\varphi^2$. In virtue of the particular form of the metric it is straightforward to see that β is the coordinate speed of light.

2.2. *The matter system*

The Yang-Mills part is represented by its vector potential $\mathscr{A}_a = A^l_a \mathbf{T}_l$, where $\mathbf{T}_l$ denotes the $\mathfrak{su}(2)$ generators which are related to the Pauli matrices as $\mathbf{T}_l = \frac{1}{2}\boldsymbol{\sigma}_l$. The corresponding field strength reads then as

$$\mathscr{F}_{ab} = 2\nabla_{[a}\mathscr{A}_{b]} + \mathrm{i}g\left[\mathscr{A}_a, \mathscr{A}_b\right], \quad F^l_{ab} = 2\nabla_{[a}A^l_{b]} - g\varepsilon_{kj}{}^l A^k_a A^j_b \tag{2}$$

where $\mathscr{F}_{ab} = F^l_{ab}\mathbf{T}_l$, while the Yang-Mills Lagrangian is given as

$$\mathfrak{L}_{YM} = -\frac{1}{2}\mathrm{Tr}\mathscr{F}_{ab}\mathscr{F}^{ab} = -\frac{1}{4}\gamma_{kj}F^k_{ab}F^{jab} \tag{3}$$

where $\gamma_{kj} = K^{-1}\mathrm{Tr}\left\langle \mathbf{T}_k, \mathbf{T}_j\right\rangle$, with $K^{-1} = 2$, is the positive definite metric on $\mathfrak{su}(2)$ which takes the simple form $\gamma_{kj} = \delta_{kj}$.

The Higgs-field is represented as $\Phi = \Phi^l\mathbf{T}_l$, with Lagrangian

$$\mathfrak{L}_H = \frac{1}{2}\left[2\mathrm{Tr}\mathfrak{D}_a\Phi\mathfrak{D}^a\Phi - V(\Phi)\right] = \frac{1}{2}\left[\gamma_{ij}\mathfrak{D}_a\Phi^i\mathfrak{D}^a\Phi^j - V(\Phi)\right] \tag{4}$$

where the gauge covariant derivative of Φ is given as

$$\mathfrak{D}_a\Phi = \nabla_a\Phi + \mathrm{i}g\left[\mathscr{A}_a, \Phi\right], \quad \mathfrak{D}_a\Phi^l = \nabla_a\Phi^l - g\varepsilon_{kj}{}^l A^k_a\Phi^j\,. \tag{5}$$

and the self-interaction potential V is assumed to possess the standard form

$$V(\Phi) = \frac{\lambda}{4}\left[\gamma_{ij}(\Phi^i\Phi^j - \Phi_0^i\Phi_0^j)\right] . \tag{6}$$

The coupled Yang-Mills–Higgs system is characterized by the Lagrangian $\mathfrak{L}_{YMH} = \mathfrak{L}_{YM} + \mathfrak{L}_H$ while its stress-energy tensor is given as

$$T_{ab} = -\frac{1}{4\pi}\left\{\mathrm{Tr}\,\mathscr{F}_{ac}\mathscr{F}_b{}^c - \mathrm{Tr}\,\mathfrak{D}_a\Phi\mathfrak{D}_b\Phi + \mathfrak{L}_{YMH}\,g_{ab}\right\} . \tag{7}$$

3. The dynamics

It can then be justified that the above specified coupled Einstein–Yang-Mills–Higgs system is governed by the field equations

$$\nabla^e\mathscr{F}_{ea} + ig\left\{[\mathscr{A}^c, \mathscr{F}_{ca}] + [\Phi, \mathfrak{D}_a\Phi]\right\} = 0, \tag{8}$$

$$\nabla_e\mathfrak{D}^e\Phi + ig\left[\mathscr{A}^c, \mathfrak{D}_c\Phi\right] + \frac{1}{2}\frac{\partial V}{\partial \Phi} = 0, \tag{9}$$

$$E^a{}_b = R^a{}_b - \frac{1}{2}g^a{}_bR - 8\pi T^a{}_b = 0\,. \tag{10}$$

Restricting our attention—as in Fodor & Rácz 2008, Fodor & Rácz 2004—to the minimal spherically symmetric configurations the Yang-Mills and Higgs fields can be given in the Abelian gauge as

$$A = -\frac{1}{g}\left[w\left(\mathbf{T}_2\mathrm{d}\vartheta - \sin\vartheta\,\mathbf{T}_1\mathrm{d}\varphi\right) + \cos\vartheta\,\mathbf{T}_3\mathrm{d}\varphi\right], \quad \Phi = H\mathbf{T}_3\,, \tag{11}$$

with $\Phi_0 = H_0\mathbf{T}_3$, for some $H_0 \in \mathbb{R}$. By introducing then $h(\tau,\rho)$ via $H(\tau,\rho) = \frac{h(\tau,\rho)}{\rho} + H_\infty$, where $H_\infty = \lim_{\rho\to\infty} H$, as a new Higgs variable the field equations (8) and (9) can be seen to take the form

$$w_{\tau\tau} = \beta^2 w_{\rho\rho} + \beta\beta_\rho w_\rho + \frac{\beta_\tau w_\tau}{\beta} + \alpha\beta^2 w\left[\frac{1-w^2}{r^2} - g^2\left(\frac{h}{\rho} + H_\infty\right)^2\right], \tag{12}$$

$$\begin{aligned} h_{\tau\tau} = \beta^2 h_{\rho\rho} + \beta\beta_\rho h_\rho + \frac{\beta_\tau h_\tau}{\beta} - \frac{2\beta^2 h_\rho + \beta\beta_\rho h}{\rho} - \frac{2r_\tau h_\tau - 2\beta^2 r_\rho\left(h_\rho - \frac{h}{\rho}\right)}{r} \\ - \alpha\beta^2\left(\frac{h}{\rho} + H_\infty\right)\left[\frac{2\rho w^2}{r^2} + \frac{\lambda h}{2}\left(\frac{h}{\rho} + 2H_\infty\right)\right] + \frac{2\beta^2 h}{\rho^2}, \end{aligned} \tag{13}$$

where the abbreviating notation $f_\sigma = \partial_\sigma f$ is applied (for $f = \alpha$, β, r, w and h).

The combinations $E^\varphi{}_\varphi + E^\vartheta{}_\vartheta - E^\rho{}_\rho - E^\tau{}_\tau = 0$, and $E^\rho{}_\rho + E^\tau{}_\tau = 0$ yield similar wave equations for α and r, while β remains freely specifiable. However, as in Csizmadia & Rácz 2010 by introducing the Misner-Sharp mass as

$$m = \frac{r}{2}\left(1 + g^{\mu\nu}\partial_\mu r\partial_\nu r\right) = \frac{r}{2}\left(1 + \frac{r_\tau^2 - \beta^2 r_\rho^2}{\alpha\beta^2}\right), \tag{14}$$

these latter equations become simpler

$$\alpha_{\tau\tau} = \underbrace{\beta^2\alpha_{\rho\rho} + \beta\beta_\rho\alpha_\rho + \frac{\beta_\tau\alpha_\tau}{\beta} + \frac{\alpha_\tau^2 - \beta^2\alpha_\rho^2}{\alpha} + 2\alpha\beta\beta_{\rho\rho} + \frac{4\alpha^2\beta^2 m}{r^3}}_{\text{geometrical part}} + \underbrace{8\pi\alpha^2\beta^2\left(T^\varphi{}_\varphi + T^\vartheta{}_\vartheta - T^\rho{}_\rho - T^\tau{}_\tau\right)}_{\text{matter part}}, \tag{15}$$

$$r_{\tau\tau} = \underbrace{\beta^2 r_{\rho\rho} + \beta\beta_\rho r_\rho + \frac{\beta_\tau r_\tau}{\beta} - \frac{2\alpha\beta^2 m}{r^2}}_{\text{geometrical part}} + \underbrace{4\pi r\alpha\beta^2 \left(T^\rho{}_\rho + T^\tau{}_\tau\right)}_{\text{matter part}}. \tag{16}$$

and, more importantly, numerically stable at the origin.

The full set of evolution equations can be put into the form

$$\partial_\tau \mathbf{u} = \mathbf{A}\partial_\rho \mathbf{u} + \mathbf{B} \tag{17}$$

with the 11-dimensional vector variable $\mathbf{u} = (m, \alpha, \alpha_\tau, \alpha_\rho, r, r_\tau, r_\rho, w, w_\tau, w_\rho, h, h_\tau, h_\rho)^T$ and a suitable source term $\mathbf{B}$, the nontrivial components of which can be determined by making use of (8), (9), (15) and (16). What is even more important the pertinent form of (17) can be seen to possess the form of a first order strongly hyperbolic system to which the existence and uniqueness of solutions is guaranteed.

In addition, by making use of the Bianchi identity, $\nabla_a E^a{}_b \equiv 0$, it can be shown that the 2-dimensional vector variable $\mathbf{u} = \left(E^\tau{}_\tau, E^\tau{}_\rho\right)^T$—built up from the left hand sides of the constraint equations $E^\tau{}_\tau = 0$ and $E^\tau{}_\rho = 0$—do satisfy a first order strongly hyperbolic system of the form (17) with a source term $\mathbf{B}$ the components of which are homogeneous and linear in $E^\tau{}_\tau$ and $E^\tau{}_\rho$. Thereby they possess identically zero solution for vanishing initial data specifications or, in other words, it is satisfactory to solve them only on the initial data surface Σ_0 as the constraints propagate. For more details see Csizmadia & I Rácz 2010 and Csizmadia & Kovács & Rácz 2010.

4. Numerical results

The first order strongly hyperbolic system of evolution equations (17) were solved numerically–as in Csizmadia & Rácz 2010–with our finite difference code called *GridRipper AMR*, within which the time integration is done by making use of the method of lines based on a fourth order Runge-Kutta scheme. *GridRipper* does also incorporate the techniques of adaptive mesh refinement Csizmadia 2007.

4.1. The initial data

The methods for the selection of suitable initial data on Σ_0 are exactly the same as in Csizmadia & Rácz 2010, with obvious changes associated with the fact that in the present case we have the Yang-Mills and Higgs fields variables, w and h, instead of a single Klein-Gordon field investigated in Csizmadia & Rácz 2010.

For simplicity and for demonstrative purposes we shall use only the type (A) initial data specification described in details in Csizmadia & Rácz 2010. In particular, in all of the numerical simulations discussed below the choices $r = \rho$, $r_\tau = 0$,

$$w = \frac{gC\rho}{\sinh(gC\rho)}, \quad h = C\rho\left[\operatorname{ctgh}(gC\rho) - 1 - \frac{1}{gC\rho}\right], \tag{18}$$

and $w_\tau = h_\tau = 0$ were made on Σ_0, with parameter values $C = H_\infty = 0.5$, $g = 0.05$ and $\lambda = 0$.

4.2. Time development

Notice first that in (18) the functional form of w and h was chosen to possess the form that of the t'Hoft-Polyakov magnetic monopole which is known to be a static solution to the Yang-Mills–Higgs equations on the flat Minkowski background. However, in the considered case the geometry is also dynamical and as it is indicated by Figs. 1 and 2 the geometry differs from that

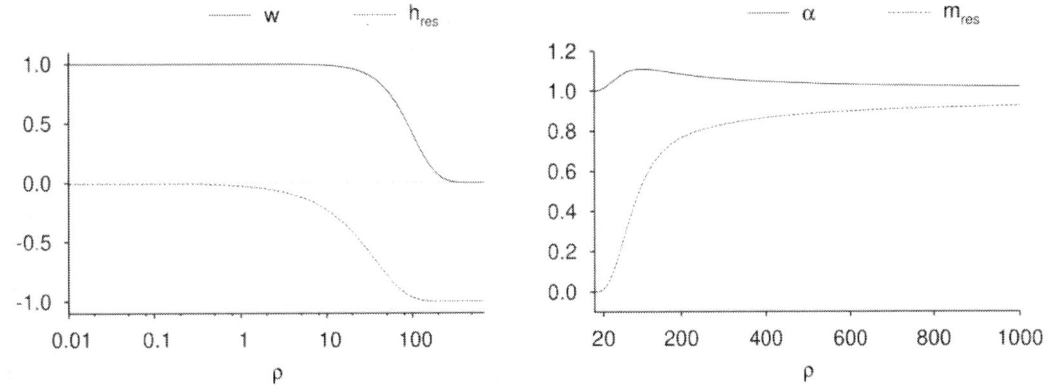

Figure 1. The functions w and α, along with the rescailed Higgs variable $h_{\rm res} = h/20$ and the rescailed Misner-Sharp mass $m_{\rm res} = m/10$, are shown on the initial data surface.

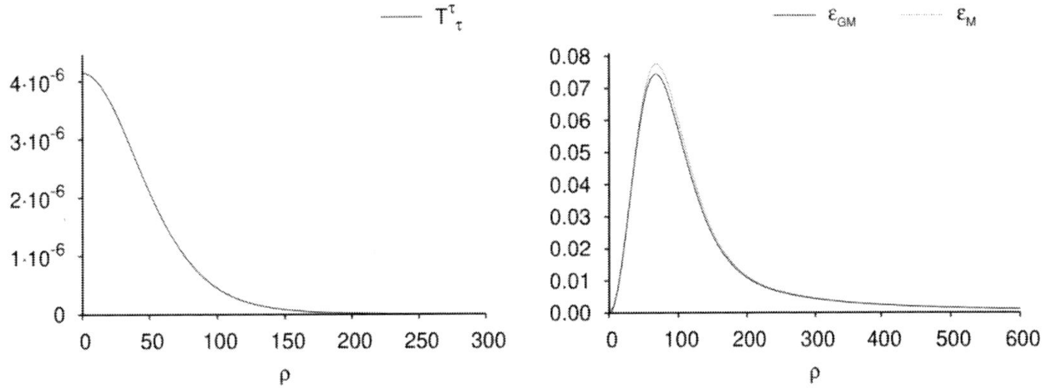

Figure 2. On the left the matter energy density $T^\tau{}_\tau$ while on the right the "weighted" energy density of matter $\mathscr{E}_M = 4\pi r^2\sqrt{\alpha}\,T^\tau{}_\tau$ and the weighted combined energy density of gravity and matter $\mathscr{E}_{GM} = 4\pi r^2(r_\rho T^\tau{}_\tau - r_\tau T^\tau{}_\rho)$ are indicated on Σ_0.

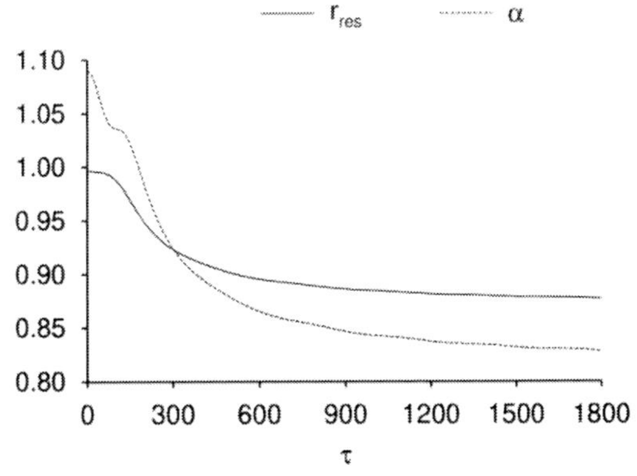

Figure 3. Time evolution of the functions α and the rescailed $r_{\rm res} = r/70$ are shown at $\rho = 69.7388$. Although there is a slight oscillation on the graph of α apparently the breathing state of the system is dominated by that of the matter.

of the Minkowski spacetime. Correspondingly, a dynamical process starts as it is indicated by Figs. 3 and 4. It is also clearly visible that a breather type oscillation develops which, in virtue of Fig. 3 is mainly dominated by the excitations of the matter system.

On Fig. 4 the weighted matter and combined gravity and matter energy densities $\mathscr{E}_M$, $\mathscr{E}_{GM}$

Journal of Physics: Conference Series **218** (2010) 012008 doi:10.1088/1742-6596/218/1/012008

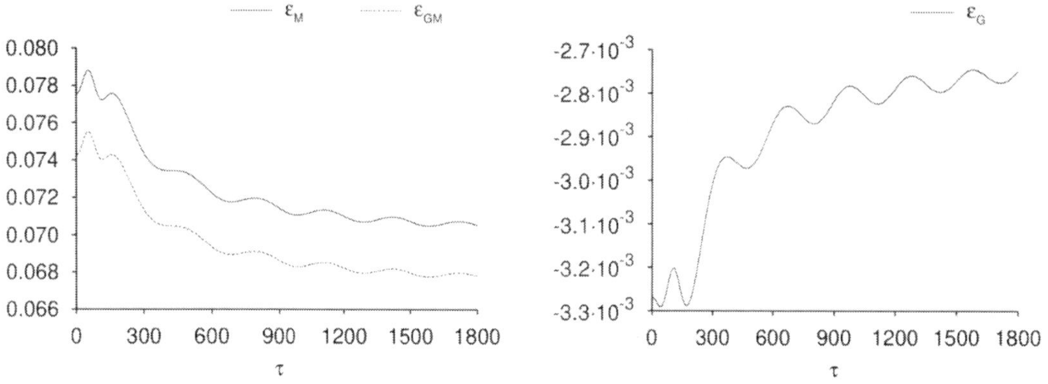

Figure 4. The time evolution of the energy density of matter $\mathcal{E}_M$, the combined energy density of gravity and matter $\mathcal{E}_{GM}$ and the binding energy density of gravity $\mathcal{E}_G$ are indicated at $\rho = 69.7388$.

(and their difference, the binding gravitational energy density, $\mathcal{E}_G = \mathcal{E}_{GM} - \mathcal{E}_M$*)* are defined such that the energy of matter, $E_M(\rho)$, or gravity and matter, $E_{GM}(\rho)$, within a ball of radius ρ can be given in terms of them given as $E_M(\rho) = 4\pi \int_0^\rho \mathcal{E}_M \, \mathrm{d}\rho$ and $E_{GM}(\rho) = 4\pi \int_0^\rho \mathcal{E}_{GM} \, \mathrm{d}\rho$.

Acknowledgments
This research was supported in part by OTKA grant K67942.

References
Bartnik R, McKinnon J 1988 *Phys. Rev. Lett.* **61** 2
Bizoń P, Chmaj T, Rostworowski A, Zajac S 2009 *Class. Quantum Grav.* **26** 225015
Bizoń P, Zenginoglu A 2009 *Nonlinearity* **22** 2473-2485
Bizoń P 1990 *Phys. Rev. Lett.* **64** 24
Breitenlohner P, Forgács P, Maison D 1994 *Commun. Math. Phys.* **163** 141-172
Csizmadia P, Rácz I 2010 *Class. Quantum Grav.* **27** 1
Csizmadia P 2007 *Class. Quantum Grav.* **24** 12
Csizmadia P, Kovács G, Rácz I 2010 *in preparation*
Fodor Gy, Rácz I 2008 *Phys. Rev.* D **77** 025019
Fodor Gy, Rácz I 2004 *Phys. Rev. Lett.* **92** 15
Forgács P, Manton N S 1980 *Commun. Math. Phys.* **72** 15-35
Goddard P, Di Olive 1978 *Rep. Prog. Phys.* **41**
Pürrer M, Aichelburg P C 2009 *Class. Quantum Grav.* **26** 3
Rácz I 2005 *Class. Quantum Grav.* **23** 1
Volkov M S, Gal'tsov D V 1999 *Phys. Rep.* **319** 1-83
Yang C N, Mills R L 1954 *Phys. Rev.* **96** 1
Zenginoglu A 2008 *Class. Quantum Grav.* **25** 17

5th Workshop of Young Researchers in Astronomy & Astrophysics IOP Publishing
Journal of Physics: Conference Series **218** (2010) 012009 doi:10.1088/1742-6596/218/1/012009

A new definition for parameterizing perturbed Kepler dynamics

János Majár

KFKI Research Institute for Particle and Nuclear Physics, Budapest 114, P.O.Box 49, H-1525 Hungary

E-mail: `majar@rmki.kfki.hu`

Abstract. Event rate estimations show that parabolic and hyperbolic encounters of compact objects will become significant sources for advanced ground-based gravitational wave detectors. To analytically describe the motion of the objects and the emitted gravitational waveforms a suitable parameterization is needed to be introduced. In the following paper such a parameterization is defined which is valid for both bound and open orbits. First post-Newtonian and spin-orbit corrections are described in detail.

1. Introduction

Event rate estimations (O'Leary at al., 2008) results that parabolic and hyperbolic encounters of compact objects provide significant sources of gravitational waves (GWs) for ground-based interferometric detectors LIGO, VIRGO, GEO, TAMA and advLIGO. A source like these provides characteristic burst-like wave signal, and can be described analitically. Furthermore, due to radiational reaction open orbits may become closed. In this case the first quasi-parabolic burst-like signal further bursts periodically, until it reaches the validity region of the usual elliptic description of the binary. The characteristic behavior and the possibility to describe them analytically greatly increases the detectability of these systems.

When investigating different corrections to the evolution, similarly to eccentric binaries these sources are described by the post-Newtonian (PN) approximation. In the case of emitted gravitational waves this method is combined with the multipolar expansion of the post-Minkowskian approximation of the radiation field (Blanchet, 2006). Although the method is developed for elliptic and circular orbits, it can be applied for open orbits, too.

Former works (Turner, 1977) use only the Keplerian description of open orbits, and neglecting relativistic corrections and interactions. To give the appropriate description of these systems the next step is to investigate the first relativistic correction to the dynamics, and the effects of the spin-orbit interaction. The general description is well known for these corrections, but a detailed investigation is still under construction.

Since the perturbed Kepler motion cannot be investigated fully analytically, an appropriate parameterization of the orbit is needed to be introduced. This question is highly detailed in many works for closed orbits (Keresztes et al., 2005) and (Damour and Deruelle, 1985). Furthermore, there are some works introducing parameterizations for 1 PN corrections in the case of open orbits, too. An other important line of the investigations is the description of the energy and angular momentum losses for eccentric and hyperbolic orbits (Peters and Mathews, 1963), (Peters, 1964) and (Hansen, 1972).

When investigating spin interactions there arise a freedom in defining the spin vector, or, equivalently the center of mass of the rotating body. This gauge freedom is treated with gauge fixing properties called spin supplementary conditions (SSC) (Pirani, 1956), (Dixon, 1964), (Newton and Wigner, 1949) and (Corinaldesi and Papapetrou, 1951). These SSC conditions are physically equivalent, but the explicit evaluations highly depend on the choice. In the following the SSC choice in (Gergely et al., 1998) is used.

In the following paper a generally applicable parameterization is defined in a way independent on the type of the orbit. Most true anomalous parameterizations are defined by the orbital angle. This is not straightforwardly applicable for spin-orbit coupling, where the orbital plane precess. The following definition does not depend on the exact notion of the "orbital angle". Besides the general method given for linear corrections first post-Newtonian and spin-orbit terms are detailed. The general description of the motion, and the radial and angular equations of motion can be found in (Majár and Vasúth, 2008).

Throughout the paper the $G = c = 1$ units are used, and δ denotes the different perturbative corrections to the expressions in a general way.

2. General definition for the parameterization of the orbit

The main goal of introducing different parameterizations is to solve the radial equation of the perturbed Keplerian motion

$$\dot{r}^2 = \frac{2E}{\mu} + \frac{2m}{r} - \frac{L^2}{\mu^2 r^2} + \delta\left(\dot{r}^2\right) . \tag{1}$$

The total energy E and the magnitude L of the orbital angular momentum are constants of motion which follows from a Lagrangian description, and $m = m_1 + m_2$ and $\mu = m_1 m_2/m$ denotes the total and relative mass of the binary. The turning points, defined by $\dot{r}^2 = 0$, are solved as

$$r_{\substack{max \\ min}} = \frac{m\mu \pm A_0}{-2E} + \delta r_{\substack{max \\ min}} , \tag{2}$$

where A_0 denotes the magnitude of the Laplace-Runge-Lenz vector ($A_0^2 = m^2\mu^2 + 2EL^2/\mu$). To introduce the radial parameterization first we investigate the following integral:

$$\tilde{\Upsilon}(r') = \Upsilon(r)|_{r_{min}}^{r'} = \int_{r_{min}}^{r'} \frac{\dot{\Upsilon} dr}{\dot{r}} ,$$

where the equation for the polar angle Υ can be evaluated following the description of the angular evolution in (Majár and Vasúth, 2008).

It is straightforward to evaluate $\tilde{\Upsilon}(r_{min}) = 0$. In the case of unbounded orbits r_{min} is the only positive root of the equation, and $r_{max} = \infty$. For elliptic orbits r_{min} and r_{max} are the turning points of the radial motion. With the use of these facts we have

$$\tilde{\Upsilon}(r_{max}) = \chi_{max} + \delta\Upsilon ,$$

where in the case of elliptic and parabolic orbits $\chi_{max} = \pi$, and for hyperbolic orbits $\chi_{max} = \arccos(-m\mu/A_0)$. Furthermore, χ_{max} is the maximum value of χ for open orbits in the Newtonian case.

With the use of the above quantity the new definition of the parameterization can be given as

$$\frac{dr}{d(\cos\chi)} = -\gamma r^2 , \qquad \chi_{\substack{max \\ min}} = \tilde{\Upsilon}(r_{\substack{max \\ min}})/K , \tag{3}$$

where K is one plus the periastron precession for the orbit.

The general solution of the radial motion for all types of orbits is

$$r(\chi)=\frac{L^2}{\mu(\mu m+A_0\cos\chi)}+\delta r\ , \tag{4}$$

and the relation between the generalized true anomaly parameter and the coordinate time is given by

$$\frac{dt}{d\chi}=\frac{1}{\dot r}\frac{dr}{d\chi}=\frac{\mu r^2}{L}+\delta\left(\frac{dt}{d\chi}\right)\ . \tag{5}$$

It can be shown that this parameterization is equivalent to the results of Damour and Deruelle for 1 PN corrections, and to the generalized true anomaly parameterization in (Keresztes et al., 2005).

The type of orbital motion, namely elliptic, parabolic and hyperbolic, is determined by the value of the total energy E. Although the formal expressions of $r(\chi)$ and $dt/d\chi$ are identical for all cases, the numerical integration of $\chi(t)$ and the determination of $r(t)$ lead to different results, which show the different behavior of the three types of orbit.

3. First post-Newtonian and Spin-Orbit corrections

The 1 PN correction for the radial equation of motion Eq.(1) is

$$\delta\left(\dot r^2\right)_{PN}=3(3\eta-1)\frac{E^2}{\mu^2}+2(7\eta-6)\frac{Em}{\mu r}-2(3\eta-1)\frac{EL^2}{\mu^3r^2}+(5\eta-10)\frac{m^2}{r^2}-(3\eta-8)\frac{mL^2}{\mu^2r^3}\ ,$$

where $\eta=\mu/m$, and the spin-orbit correction is

$$\delta\left(\dot r^2\right)_{SO}=\frac{2E\mathbf{L}\boldsymbol{\sigma}}{m\mu^2r^2}-\frac{2}{\mu r^3}(2\mathbf{LS}+\mathbf{L}\boldsymbol{\sigma})\ ,$$

where $\mathbf{S}=\mathbf{S}_1+\mathbf{S}_2$ and $\sigma=\mathbf{S}_1m_2/m_1+\mathbf{S}_2m_1/m_2$. Henceforth the corrections for the $\delta r_{\substack{max\\min}}$ turning points are

$$\begin{aligned}\delta r^{PN}_{\substack{max\\min}} &= (\eta-7)\frac{m}{4}\pm(\eta+9)\frac{m^2\mu}{8A_0}\mp(3\eta-1)\frac{A_0}{8\mu}\ ,\\ \delta r^{SO}_{\substack{max\\min}} &= \frac{2\mathbf{LS}+\mathbf{L}\boldsymbol{\sigma}}{L^2A_0}(A_0\mp m\mu)\pm\frac{E\mathbf{L}\boldsymbol{\sigma}}{A_0m\mu}\ .\end{aligned}$$

The most important step when determining the perturbative terms of the parameterization is to evaluate the periastron precession corrections at 1 PN order and for the spin-orbit interaction. The result is

$$K=\left(1+\frac{3m^2\mu^2}{L^2}-\frac{m\mu^3(3\mathbf{L}\boldsymbol{\sigma}+4\mathbf{LS})}{L^4}\right)\ . \tag{6}$$

With the use of the above definition the corrections to $r(\chi)$ are evaluated as

$$\begin{aligned}\delta r_{PN} &= -\frac{2(6-\eta)m^4\mu^6+2(10-3\eta)EL^2m^2\mu^3+(1-3\eta)E^2L^4}{2A_0\mu^3(\mu m+A_0\cos\chi)^2}\cos\chi\\ &- \frac{2(2-\eta)mEL^2+(6-\eta)m^3\mu^3}{\mu(\mu m+A_0\cos\chi)^2}\ ,\\ \delta r_{SO} &= \frac{2(2\mathbf{LS}+\mathbf{L}\boldsymbol{\sigma})\Big[A_0(2m^2\mu^3+EL^2)+m\mu(2m^2\mu^3+3EL^2)\cos\chi\Big]}{A_0L^2(\mu m+A_0\cos\chi)^2}\\ &- \frac{2E\mathbf{L}\boldsymbol{\sigma}\Big[A_0m\mu^2+(m^2\mu^3+EL^2)\cos\chi\Big]}{A_0m\mu^2(\mu m+A_0\cos\chi)^2}\ .\end{aligned}$$

Again, the above expressions are valid for all the three types of orbits. To complete the evaluations the 1 PN order and spin-orbit terms in the Eq.(5) differential equations become

$$\left(\frac{dt}{d\chi}\right)_{PN} = -\frac{\mu r^2}{2L^3}\left[(\eta-13)m^2\mu^2+(3\eta-1)A_0^2+(3\eta-8)m\mu A_0\cos\chi\right] ,$$

$$\left(\frac{dt}{d\chi}\right)_{SO} = -\frac{\mu r^2}{mL^5}\left[(2\mathbf{LS}+\mathbf{L}\boldsymbol{\sigma})m\mu^2(3m\mu+A_0\cos\chi)-EL^2\mathbf{L}\boldsymbol{\sigma}\right] .$$

4. Detectable gravitational waveform

As it has been detailed the above definition gives rise to expressions formally independent of the value of the energy, or, equivalently of the type of the orbit. Since all the methods used to evaluate the detectable waveform are independent of it too, the $h(\chi)$ expressions are formally the same for elliptic, parabolic and hyperbolic cases.

When generating realistic gravitational waveform templates (Kidder, 1995) and (Will and Wiseman, 1996) the numerical integration of the inverse of Eq.(5) is needed. During this procedure there arise three completely different results for elliptic, parabolic and hyperbolic orbits. These differences leads to highly different gravitational wave signals, too. In the case of elliptic orbits the expected waveform signals are periodic (or, including radiational reaction, quasi-periodic), in the case of hyperbolic orbits the signal is burst-like, and in the case of parabolic orbits the burst-like signal evolves into a quasi-periodic waveform if the radiational reaction is included in the description.

The above definition gives rise to the opportunity to determine the higher order corrections to the parameterization of the orbit in a way independent of the type of the perturbed Keplerian orbit. Henceforth, the corrections of further effects (namely, quadruple-monopole, magnetic dipole-dipole interactions, etc.) can be easily evaluated. This method provides the possibility to give the most realistic theoretical waveform signals for the gravitational wave detector template banks.

Acknowledgments

This work was supported by OTKA grant No. NI68228.

References

Blanchet L 2006 *Living Rev. Relativity* **9** 4
Corinaldesi E and Papapetrou A 1951 *Proc. Roy. Soc. London* A **209** 259
Damour T and Deruelle N 1985 *Ann. Inst. Henri Poincaré* **43** 107
Dixon W 1964 *Nuovo Cim.* **34** 317
Gergely L Á, Perjés Z and Vasúth M 1998 *Phys. Rev.* D **58** 124001
Hansen R O 1972 *Phys. Rev.* D **5** 1021
Keresztes Z, Mikóczi B and Gergely L Á 2005 *Phys. Rev.* D **72** 104022
Kidder L E 1995 *Phys. Rev.* D **52** 821
Majár J and Vasúth M 2008 *Phys. Rev.* D **77** 104005
Newton T D and Wigner E P 1949 *Rev. Mod. Phys.* **21** 400
O'Leary R M, Kocsis B and Loed A *arXiv*:0807.2638v2[astro-ph]
Peters P C and Mathews J 1963 *Phys. Rev.* **131** 435
Peters P C 1964 *Phys. Rev.* **136** B1224
Pirani F A E 1956 *Acta Phys. Polon.* **15** 389
Turner M 1997 *Astrophys. J.* **216** 610
Will C M and Wiseman A G 1996 *Phys. Rev.* D **54** 4813

5th Workshop of Young Researchers in Astronomy & Astrophysics IOP Publishing
Journal of Physics: Conference Series **218** (2010) 012010 doi:10.1088/1742-6596/218/1/012010

Compact Stars in Kaluza – Klein World

Gergely Gábor Barnaföldi, Péter Lévai, Béla Lukács
KFKI RMKI of the HAS, 29-33 Konkoly-Thege Miklós Str. H-1121, Budapest Hungary
E-mail: bgergely@rmki.kfki.hu

Abstract. Unification and geometrization of interactions has been extensively studied during the XX. century. In this short contribution we investigated the possible effect of an extra compactified dimension (alias hypercharge) on a flavor dependent gravitational potential, proposed by Fischbach et al.. We estimated the deviation from the 3+1 dimensional scheme and found that, although the deviation is moderate, for celestial compact object it may be higher by orders of magnitude than in terrestrial laboratory measurements.

1. Introduction

Searching for signatures of compactified extra dimension(s) is an interesting and challenging task. There are no direct observables for these topological issues in the classical physical cases, so search can start at highest energy systems, which can be created in particle accelerators or existed inside celestial objects as compact stars or black holes (Barnaföldi *et al.*, 2006).

In this short review we would like to re-think and extend the usual description of compact objects in case of more than $3+1$ dimensional space-time – in a Kaluza – Klein-like world with a compactified extra dimension 1_c. Our aim is to highlight a possible deviation from standard gravitational potential as an influence of compactified extra dimension (or hypercharge) in a compact pulsar binary (like e.g. PSR J0737-3039).

2. Geometrization of Interactions – a Historical Review

Einstein's general relativity (GR) was the first theory which translated an interaction of the nature i.e. gravity into a geometrical approach. The original theory was developed in 1916, assumed to have symmetric metric tensor in $3+1$ space-time.

Later Einstein worked more on the development of the unification of gravity and electromagnetic interaction (Cartan 1923). The generalization of his model gave up the symmetric being of the metric tensor which leads to 16 free parameters in $3+1$ dimensional space-time. In case of a non-symmetric metric tensor, gravity requires 10 independent parameter and 6 were trivially left for the Maxwell tensor. Even if this generalization looks trivial, it leads to several non-physical consequences.

Another direction was proposed by Kaluza (1921) and later on extended by Klein (1926), where they kept the assumption of the general relativity using the symmetric metric tensor, but they extended the $3+1$ dimensional space-time including a spatial compactified extra 5th dimension. In this case the number of independent parameters become 15, separated as: 10 for gravity, 4 correspond to the vector potential, and finally setting $g_{55} = 1$ is crucial to fix the remaining part. However, this latter assumption violates the role of the '*free choice of coordinate systems*', but Maxwell equations can be carry out in a weak-field approximation for any case.

Brans and Dicke (1961) further developed the theory introducing a time dependent 'gravitational constant', $G(t)$, giving an additional equation based on cosmological issues. In this way the whole gravitational and electro-magnetic theory can be unified and geometrized. Following the this second scenario, Chodos and Detweiler (1980) did the separation of the 5th direction for the first time in a consequent way in $4+1$ dimensional space-time. They find the 15 free 'geometrical parameters' can be separated to $10+4+1$ components, where the last 1 dimensional part must be a scalar field.

Lukács and Pacher (1985) reviewed the Brans – Dicke theory, and pointed out: the generation of extra mass by fast moving in the 5th direction is limited due to the charge-to-mass ratio of the observed particles such as the proton. This model ruled out the electro-magnetic origin of the charge. On the other hand, applying the model for the weak charge, opened a new directions showing the possibility for the unification and geometrization of the weak and gravitational theories. In this picture the extra compactified 5th dimension represents the hypercharge or strangeness.

3. On the Shadow of a Dimension

Obviously, a microscopical and compactified dimension has 4-dimensional consequences. However, these consequences may depend on the nature of higher dimensions, so our present study cannot give a full review. For now we use some simplifying conditions, rather accepted in general relativity but not always in particle physics. They are as follows:

(i) The space-time has $(3+m_c)+1$ dimensions; and except for the last one, they are space-like, while the last is time-like.

(ii) The structure of general relativity is just as we have learnt in $3+1$ dimensions; especially the form of the *Equivalence Principle* is unchanged.

(iii) All causality postulates, including lightcone structure, are as they were in the $3+1$ case.

(iv) The m_c extra space-like dimensions are microscopical, i.e. they are all compact with microscopical circumferences.

(v) There is complete Killing symmetry in the m_c-dimensional microscopical subspace.

(vi) In this contribution, we focus on the simplest case, thus for now: $m_c = 1$.

Cond. (i) excludes inherently acausal space-times – 2 simultaneous time-like coordinates would permit time travel as an everyday profession (Lukács 2003). Conds. (ii) and (iii) are natural in general relativity contrasted to particle physical theories where the extra dimensions are 'not exactly as our familiar (iii)', e.g. that motion in that direction 'is not like other motions'. Then Cond. (iv) is for keeping the extra dimensions 'really microscopical'. The last two conditions are not really fundamental but simplifying conditions for now. Cond. (v) is for avoiding difficulties about the directly unobservable microstructures and Cond. (vi) for restricting ourselves for the simplest cases.

The general metric, satisfying Conds. (i–vi) can always be written into the form

$$\mathrm{d}s^2 = -\Phi^{2/3}(\mathrm{d}\Psi + A_r \mathrm{d}x^r)^2 + \Phi^{-1/3} g_{rs}\mathrm{d}x^r \mathrm{d}x^s \ , \tag{1}$$

with an Einstein convention for summation in 4 dimensions, where $\Psi = x^5$, and $r, s = 0, 1, 2, 3$ and the quantities Φ and A_i do not depend on Ψ. Then the Lagrangian of the 5 dimensional general relativity gets the form in 4 dimensions as

$$\mathcal{L} = R^{(5)} = R^{(4)} + \frac{\Phi}{4}\ F^{rs}F_{rs} + \frac{g^{rs}}{6} \cdot \frac{\Phi_{,r}\Phi_{,s}}{\Phi^2} \ . \tag{2}$$

So from a purely geometrized 5-dimensional space-time you get a 4-dimensional 'projection' with a scalar coupling and a vector field as well.

The scalar field corresponds to the scalar coupling in the Brans–Dicke general relativity (Brans & Dicke, 1961). The vector field has the same structure as Maxwell field, and therefore it was the feeling of Kaluza that electromagnetism has been geometrized.

However the equations of motion give some constraint for the specific charge. The motion in a 5-dimensional vacuum must happen on a 5-dimensional geodesics on one hand and on the other, equivalent with a 4-dimensional motion in the vector and scalar fields. Now the geodesic equation formally has 5 components, but the 0th one is a consequence of the normalized nature of the velocity vector,

$$u^r u_r + u^5 u_5 = +1 . \tag{3}$$

Note, here we used again the 4-dimensional convention, where, according to Conds. (i) and (iii),

$$\text{sign}(g_{ik}) = \{+, -, -, -, -\}; \quad i = 0, 1, 2, 3, 5 . \tag{4}$$

The 5th component of the geodesic equation gives a balance equation for the extra velocity component, so approximately a conservation of something seen as a charge in 4 dimensions. The larger 5th component of the velocity, indicates the larger specific charge.

However, this 5th component is not seen, so in the normalization eq. (3) we miscalculate u^0. Using the full u^i vector in the formalism we get that the specific charge goes to a finite value as $u^5 \to \infty$ (Lukács & Pacher, 1985). This problem is sometimes called the 'large mass problem'. Of course, it would not appear with large u^5 and moderate u^0 values; but such velocities would be space-like (Gegenberg & Kunstätter, 1985). To permit such velocities is contrary to Cond. (iii), the causality postulates in 5 dimensions.

Finally, we remain at a vector 'interaction' (geometrized in 5 dimensions but not in 4) with a coupling constant or specific charge:

$$\frac{q}{m^{(4)}} < \sqrt{16\pi G} \tag{5}$$

Let's take for example a proton, p, with 4 dimensional mass $m^{(4)} = m_p$ and a charge as $e = q$, which results: $e^2/(Gm_p^2) = 1.2 \times 10^{36}$. Based on eq. (5) this interaction cannot be the electromagnetic one, but much weaker vector fields can be the 'shadows' of the 5th dimension.

Applying quantum mechanics in 5 dimensions, the compactified extra dimension results in a quantized charge (Souriau, 1963) $q \simeq 2\hbar\sqrt{G}/(cR_c)$ where R_c is the compactified radius with value probably 10^{-13} cm $\leq R_c \leq 10^{-9}$ cm as we pointed out in e.g. Barnaföldi *et al.*, (2006). Note, this is well within the spatial resolution of the Large Hadron Collider (LHC) at CERN, which is $\sim \hbar c/E_{beam} \sim 3 \cdot 10^{-18}$ cm, at the highest working energies.

4. Possible Experimental Evidences

As it was shown earlier by Barnaföldi *et al.* (2006) and Kan & Shiraishi (2002) in a Kaluza–Klein like world one kind of 5 dimensional particles generate series of particles in the 4 dimensional world, following the Kaluza–Klein mass-ladder. However, it is still a question which quantum number corresponds to the mass-steps of this ladder. One can assume strangeness, S is related to the radius of the compactified extra dimension with size $\sim 10^{-13}$ cm. In this case a 5 dimensional compact baryon star would mimic a hyperon star in 4 dimensions, but with a somewhat different mass distribution, $M(R)$ (see further details in Barnaföldi *et al.*, (2006)).

In this contribution we focus on another possible signal, following the popular idea of Fischbach *et al.* from the mid '80s. They re-analyzed the Eötvös experiment about the equivalence of inertial and gravitational masses, and found a small, but systematic deviation from chemical composition of the test bodies (Eötvös *et al.*, 1922). Fischbach and collaborators find the best linear fit using hypercharge, $H = B + S$ as the source of deviation. Therefore, they believed to found the so called '5th force', which may even be responsible for the CP violation of the weak interaction (Fischbach *et al.* 1986).

Here, we do not want to argue either for or against the Fischbach's scheme however, note, their fitted coupling is $q^2/e^2 \approx 10^{-38} - 10^{-41}$, which is in agreement with eq. (5). Thus the results of Eötvös's experiment may originate from multi-dimensional general relativity.

5. Summary – Compact Stars in Kaluza–Klein World

To evaluate the above predicted effect of an extra dimension, let's use an astrophysical scenario: taking a compact binary star with typical physical parameters ($R \approx 10$ km, $M \approx 2 \times M_{\odot}$, $B \approx 2.6 \times 10^{57}$, and baryon density, $\rho_B \approx 0.15$ fm^{-3}).

Following Fischbach's idea the presence of the 5th force in 3 + 1 dimensional space-time appears as an extra term additional to the gravitational potential,

$$V(r) = -G_{\infty}\frac{m_1 m_2}{r}\left(1 + \alpha \cdot \mathrm{e}^{-r/\lambda}\right) \quad , \tag{6}$$

where G_{∞} is the gravitational constant at distance $r \to \infty$ and m_1, m_2 are the masses. Parameters, based on Fischbach's re-analysis are: $\alpha = -(7.2 \pm 3.) \times 10^{-3}$ and $\lambda = 200 \pm 50$ m.

However, the 3-dimensional effective potential is somewhat misleading not including the equivalence principle, well above 200 m distance the '5th force' cannot be felt, but eq. (6) implies that gravitating mass also contains corrections. Note, our present scheme is completely general relativistic, so the gravitational mass must be equal to the inertial one.

Obviously the maximal deviation from the '4-*dimensional equivalence principle*' would appear as using strangeness content test particles. For example if one could have Λ^0- or K^0-matter content test bodies[1] for the gravitational experiment, the difference would appear in the 2nd or 3rd digits not only at the 9th as proposed by Fischbach *et al.*, 1986.

Test bodies would feel their strangeness and in astrophysical situations hyperonic masses can occur. Especially, the '*first compact pulsar binary*', PSR J0737-3039 was discovered in 2004 by Lyne *et al.*, with $8 \cdot 10^5$ km orbit, is the best known candidate for experimental test of the extra-dimensional influence on gravitating mass in the 2nd or 3rd digit.

Acknowledgments

Acknowledgments Our work was supported in part by Hungarian OTKA T043455, T047050, PD73596, and NK62044, and supported by the János Bolyai Research Scholarship of the HAS.

References

Barnaföldi G G, Lukács B and Lévai P 2007 *Astron. Nachr.* **328** 809
Brans C and Dicke R H 1961 *Phys. Rev.* **124** 925
Cartan É 1923 *Ann. Éc. Norm. Sup.* **40** 325
Chodos A and Detweiler S 1980 *Phys. Rev. D* **21** 2167
Chodos A 1984 *Comm. Nucl. Part. Phys.* **13** 171
Eötvös von R, Pekár D and Fekete D 1922 *Annalen Phys.* **68** 11
Fischbach E *et al.* 1986 *Phys. Rev. Lett.* **56** 3
Gegenberg J and Kunstätter G 1984 *Phys. Lett. A* **106** 410
Kaluza Th 1921 *Sitzungsber Preuss. Akad. Wiss. Phys. Math Kl.* **LIV** 966
Klein O 1926 *Z. Phys.* **37** 875
Kan N and Shiraishi K 2002 *Phys. Rev D* **66** 105014
Lukács B 2003 *The Nature of Time*, eds. R. Buccheri & al. (Kluver Academic, Netherlands) pp 277
Lukács B and Pacher T 1985 *Phys. Lett. A* **113** 200
Lyne A G *et al.* 2004 *Science* **303** 1153
Souriau J-M 1963 *Nuovo. Cim.* **XXX** 565

[1] Note, mesons with zero baryon number, $B = 0$ and with non-zero strangeness (or hypercharge) $S \neq 0$, would have higher possibility to move to the 5th direction. Thus deviation effect is expected to be stronger compared to ordinary baryonic matter. On the other hand Λ^0 content compact star is more physically than K^0 content one.

5th Workshop of Young Researchers in Astronomy & Astrophysics IOP Publishing
Journal of Physics: Conference Series **218** (2010) 012011 doi:10.1088/1742-6596/218/1/012011

Elliptic waveforms for inspiralling compact binaries

Balázs Mikóczi

KFKI Research Institute for Particle and Nuclear Physics, Budapest 114, P.O.Box 49, H-1525 Hungary

E-mail: mikoczi@rmki.kfki.hu

Abstract. The inspiral of supermassive black hole binary systems with high orbital eccentricity are the most promising sources for the gravitational wave observatories. The importance of elliptic gravitational waveforms in various physical scenarios has been emphasized by several authors (Wahlquist 1987, Moreno-Garrido, Buitrago and Mediavilla 1994, Martel and Poisson 1999). Taking into account the eccentricity of the orbit in the total waveform improves the parameter estimation for these sources, as it is shown by the construction and analyzation of the Fisher information matrix. In our work we use the Fourier-Bessel analysis of the Kepler motion and the stationary phase approximation of time-depend waveforms.

1. Introduction

Compact binaries of black holes and neutron stars are important sources of gravitational radiation. The coalescence of black hole binaries, for example, will be detectable by the Laser Interferometer Space Antenna (LISA) (Bender et al 1998). The sensitivity domain for this detector is $10^{-4} - 10^{-1} Hz$ which corresponds to the inspiral of two $(10^4 - 10^7) M_{\odot}$ black holes. For circular orbits the problem of parameter estimation for LISA has been discussed in (Cutler 1998, Vecchio 2004, Berti et al 2005, Lang and Hughes 2006, Trias and Sintes 2008).

In their work Peters and Mathews have shown that the orbit of the binary are circularizing due to gravitational radiation (Peters and Mathews 1963, Peters 1964). Several authors suppose circular orbits in the gravitational wave computations. However, this circularization does not always take place. In the presence of interstellar gas the binary system can remain eccentric until the final coalescence (Armitage and Narayan 2005). N-body simulations are also confirm the relevance of the eccentricity (Preto et al 2009). Coalescing supermassive black holes in galactic nuclei with eccentricity emit detectable wave signal significantly different from those of circular orbits (O'Leary et al 2008).

The elliptic waveform is simply derived from Einstein's quadrupole formula (Thorne 1980). The first explicit elliptic waveform appears in (Wahlquist 1987, Martel and Poisson 1999) with true anomaly. Some authors have derived the waveform in time domain (Moreno-Garriodo et al 1994 and 1995, Pierro et al 2001). The parameter estimation for eccentric binaries is discussed by (Seto 2001, Barack and Cutler 2004, Yagi and Tanaka 2009).

In this paper we demonstrate that the eccentricity of orbit is important in some physically relevant situations. We use elliptic waveforms for parameter estimation of $\left(10^6 - 10^6\right) M_{\odot}$ massive black hole binaries and point out the differences between circular and elliptic orbits. We introduce geometric units $G = c = 1$ and the I, II notations denote two independent detectors of LISA.

2. Waveform of the eccentric binary

Consider a system of two point particles in an Keplerian elliptic orbit. The parametrization of the radial motion is

$$r = a(1-e^2)/(1+e\cos\phi) \tag{1}$$

with the semimajor axis a, the eccentricity e and the true anomaly ϕ of the Keplerian orbit. According to the quadrupole formula of the gravitational radiation the two polarization states in the transverse-traceless (TT) gauge are written as (Wahlquist 1987):

$$\begin{aligned} h_\times(\phi) &= -\frac{\mu m \cos\Theta}{a(1-e^2)D_L}\left[(5e\sin\phi + 4\sin 2\phi + e\sin 3\phi)\cos 2\gamma \right. \\ &\left. - \left(5e\cos\phi + 4\cos 2\phi + e\cos 3\phi + 2e^2\right)\sin 2\gamma\right] , \qquad (2) \\ h_+(\phi) &= -\frac{\mu m\left(1+\cos^2\Theta\right)}{a(1-e^2)D_L}\left[\left(\frac{5e}{2}\cos\phi + 2\cos 2\phi + \frac{e}{2}\cos 3\phi + e^2\right)\cos 2\gamma \right. \\ &\left. + \left(\frac{5e}{2}\sin\phi + 2\sin 2\phi + \frac{e}{2}\sin 3\phi\right)\sin 2\gamma + \left(e\cos\phi + e^2\right)\frac{\sin^2\Theta}{1+\cos^2\Theta}\right] , \qquad (3) \end{aligned}$$

where D_L is the luminosity distance, Θ is the direction of the angular momentum (inclination), γ is the angle of the periastron position, $m = m_1 + m_2$ is the total mass and $\mu = m_1 m_2/m$ is the reduced mass of the system. The time dependence can be computed with the use of the Fourier-Bessel decomposition. From Eqs. (2) and (3) the polarization modes can be derived (Eqs. (17) and (18) of (Moreno-Garriodo et al 1995))

$$\begin{aligned} h_\times(t) &= -h\cos\Theta\sum_n\left[\frac{S_n - C_n}{2}\sin(nl + 2\gamma) + \frac{S_n + C_n}{2}\sin(nl - 2\gamma)\right] , \qquad (4) \\ h_+(t) &= -\frac{h}{2}\sum_n\left[\sin^2\Theta A_n\cos nl + \left(1+\cos^2\Theta\right)\right. \\ &\left. \times\left(-\frac{S_n - C_n}{2}\cos(nl + 2\gamma) + \frac{S_n + C_n}{2}\cos(nl - 2\gamma)\right)\right] , \qquad (5) \end{aligned}$$

with $h = 4\mu m/aD_L$ and

$$\begin{aligned} S_n &= -\frac{2\left(1-e^2\right)^{1/2}}{e}n^{-1}J_n'(ne) + \frac{2\left(1-e^2\right)^{3/2}}{e^2}nJ_n(ne) , \qquad (6) \\ C_n &= -\frac{2-e^2}{e^2}J_n(ne) + \frac{2\left(1-e^2\right)}{e}J_n'(ne) , \qquad A_n = J_n(ne) , \qquad (7) \end{aligned}$$

here $J_n'(ne) = n[J_{n-1}(ne) - J_{n+1}(ne)]/2$ and l is the mean anomaly defined by the Kepler equation

$$l = \xi - e\sin\xi = 2\pi\nu(t-\tau) , \tag{8}$$

where ξ is the eccentric anomaly and ν is the Keplerian orbital frequency (thereafter we set $\tau = 0$). In the circular limit ($e \to 0$) the elliptic waveforms (2), (3) reduce to

$$h_\times^c(t) = -h_c\cos\Theta\sin\psi , \qquad h_+^c(t) = \frac{h_c}{2}\left(1+\cos^2\Theta\right)\cos\psi , \tag{9}$$

with the phase of the circular orbit $\psi = 2\pi\int(f/\dot{f})df$ and $h_c = 4\mu(m\pi f)^{2/3}/D_L$ (using the expressions $(2\pi\nu)^3 a^3 = m$ (Kepler's third law) and $f = 2\nu$ (Peters and Mathews 1963)).

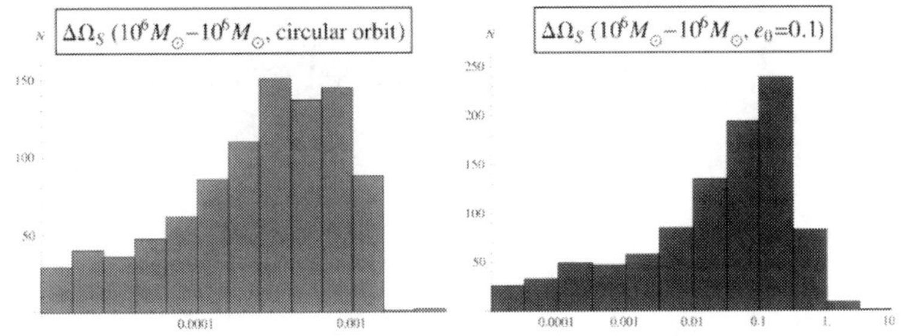

Figure 1. Distribution of the angular resolution $\Delta\Omega_S$ for circular (left) and eccentric (right) waveforms. The results are valid for 1-year observation of $(10^6 - 10^6)\, M_\odot$ black hole binaries at $z = 1$ ($D_L = 6.4Gpc$) by LISA. We have used initial position according to (Cutler 1998) and neglected the Doppler phase. The initial eccentricity is $e_0 = 0.1$ in the eccentric case (the evolution of the eccentricity due to gravitation radiation was given in (Peters 1964) to the leading order as $e(\nu) = e_0\,(\nu/\nu_0)^{-19/18}$, where ν_0 is the initial frequency, the frequency range is the same in both cases). The histograms show the results of two Monte Carlo simulations with 10^3 sources randomly located in the sky.

3. Fourier transformation of the waveform

The frequency dependence of the waveform can be computed with its Fourier transformation in the stationary phase approximation (SPA) (Moreno-Garriodo et al 1994): if $g(t)$ is an arbitrary function of time, then $\int_{-\infty}^{\infty} h[\nu(t)]e^{ig(t)}dt = \sqrt{\frac{2\pi}{\ddot{g}(t_0)}}h(\nu)]e^{i(g(t_0)+\pi/4)}$, where $\nu = \nu(t_0)$ is the time of the sampling point at the given frequency. We ignore the periastron advance [$\dot{\gamma} = 0$ in Eqs. (4) and (5)] and the Doppler phase. In this case the polarization modes are ($\gamma = 0$):

$$h_\times(\nu)_n = -\frac{\mathcal{M}^{5/6}(2\nu)^{-7/6}}{\pi^{2/3}D_L}k(n,e)\cos\Theta S_n e^{i\Psi_+}\,, \tag{10}$$

$$h_+(\nu)_n = -\frac{\mathcal{M}^{5/6}(2\nu)^{-7/6}}{2\pi^{2/3}D_L}k(n,e)\left[\sin^2\Theta A_n + \left(1+\cos^2\Theta\right)C_n\right]e^{i\Psi_-}\,, \tag{11}$$

where $\mathcal{M} = \mu^{3/5}m^{2/5}$ is the chirp mass, $\Psi_\pm = 2\pi n\nu t(\nu) - \Phi(\nu) \pm \pi/4$ (where $t(\nu)$ and $\Phi(\nu)$ can be computed from the evolution of orbital frequency due to gravitational radiation (Yunes et al 2009)) is the "eccentric phasing function" (for circular orbits the phasing is $\Psi = 2\pi f t_c - \phi_c - \pi/4 + 3/4(8\pi\mathcal{M}f_c)^{-5/3}$, where t_c and ϕ_c is the coalescing time and phase) and

$$k(n,e) = \sqrt{\frac{5(1-e^2)^{7/2}}{48n\left(1+\frac{73}{24}e^2+\frac{37}{96}e^4\right)}}\,. \tag{12}$$

Then the frequency domain waveforms detected by the I. and II. LISA detectors are

$$h^{I,II}(\nu) = \frac{\sqrt{3}}{2}\sum_n\left[F_\times^{I,II}h_\times(\nu)_n + F_+^{I,II}h_+(\nu)_n\right]\,, \tag{13}$$

where the antenna-beam pattern for the LISA was given in (Cutler 1998). The $F_{\times,+}^{I,II}$ antenna-beam pattern are rather complicated and time dependent.

4. Parameter estimation

The definition of the Fisher information matrix is

$$\mathbf{\Gamma}_{ab} = 2\sum_{I,II} \int_{\nu_{\min}}^{\nu_{\max}} \frac{\frac{\partial h^{I,II}(\nu)}{\partial\lambda^a}\frac{\partial \tilde{h}^{I,II}(\nu)}{\partial\lambda^b} + \frac{\partial \tilde{h}^{I,II}(\nu)}{\partial\lambda^a}\frac{\partial h^{I,II}(\nu)}{\partial\lambda^b}}{S_h(\nu)} d\nu \,, \tag{14}$$

where $\tilde{h}$ is the complex conjugate of h, $S_h(\nu)$ is the noise spectra of the detector (we used the LISA sensitivity data file) and λ_a are the parameters of the waveform. The correlation matrix Σ, which is the inverse of the Fisher matrix, $\Sigma = (\Gamma)^{-1}$, defines the parameter errors: $\Delta\lambda_a = \sqrt{\mathbf{\Sigma}_{aa}}$. The error of the angular resolution is

$$\Delta\Omega_S = 2\pi\sqrt{(\Delta\mu_S\Delta\phi_S)^2 - \langle\delta\mu_S\delta\phi_S\rangle^2} \,, \tag{15}$$

with $\mu_S = \cos\theta_S$, $\langle\delta\lambda_a\delta\lambda_b\rangle = \mathbf{\Sigma}_{ab}$ and the location of the binary source (μ_S, ϕ_S). For the estimation of the parameter errors we have considered $\left(10^6 - 10^6\right) M_\odot$ black holes binaries with $e_0 = 0.1$. Our frequency domain results are in agreement with those in (Cutler 1998) (1-year observation up to the innermost stable circular orbit). We have used 9 parameters in the elliptic case: $\mathcal{M}$, D_L, t_c, ϕ_c, μ_S, ϕ_S, μ_L, ϕ_L and e_0.

5. Summary

In this work we have analyzed eccentric waveforms of Keplerian orbits. During the calculations the periastron advance and the Doppler phase modulation in the phasing were neglected. We have presented how the results of parameter estimation change due to the eccentricity by computing the angular resolutions for 10^3 binaries with Monte Carlo simulation (Fig. 1).

6. Acknowledgment

This work was supported by the Hungarian Scientific Research Fund (OTKA) grant No. NI68228. I would like to thank Mátyás Vasúth and Péter Forgács for their help during my work.

7. References

Armitage P J, Natarajan P 2005 *Astrophys. J.* **632** 921
Barack L and Cutler C 2004 *Phys. Rev. D* **69** 082005
Bender P L et al. 1998 *LISA Pre-Phase A Report* 2nd ed. See http://lisa.jpl.nasa.gov
Berti E, Buonanno A, and Will C M 2005 *Phys. Rev. D* **71** 084025
Cutler C 1998 *Phys. Rev. D* **57** 7089
Lang R and Hughes S A 2006 *Phys. Rev. D* **74** 122001
Martel K and Poisson E 1999 *Phys. Rev. D* **60** 124008
Moreno-Garrido C, Mediavilla E and Buitrago J 1995 *Mon. Not. R. Astron. Soc.* **274** 115
Moreno-Garrido C, Buitrago J, and Mediavilla E 1994 *Mon. Not. R. Astron. Soc.* **266** 16
O'Leary R M, Kocsis B and Loeb A 2008 *Mon. Not. R. Astron. Soc.* **395** 2127
Peters P C and Mathews S 1963 *Phys. Rev.* **131** 435
Peters P C 1964 *Phys. Rev.* **136** 1224
Pierro V, Pinto I M, Spallicci A D, Laserra E and Recano F 2001 *Mon. Not. R. Astron. Soc.* **325** 358
Preto M, Berentzen I, Berczik P, Merritt D and Spurzem R 2009 *J. Phys.:Conf. Ser.* **154** 012049
Seto N 2001 *Phys. Rev. Lett.* **87** 251101
Thorne K S 1980 *Rev. Mod. Phys.* **52** 299
Trias M and Sintes A M 2008 *Phys. Rev. D* **77** 024030
Yagi K and Tanaka T 2009 *Constraining alternative theories of gravity by gravitational waves from precessing eccentric compact binaries with LISA* gr-qc/0906.4269
Yunes N, Arun K G, Berti E and Will C M 2009 *Phys. Rev. D* **80** 084001
Vecchio A 2004 *Phys. Rev. D* **70** 042001
Wahlquist H 1987 *Gen. Rel. Grav.* **19** 1101
http://www.srl.caltech.edu/~shane/sensitivity/

5th Workshop of Young Researchers in Astronomy & Astrophysics IOP Publishing
Journal of Physics: Conference Series **218** (2010) 012012 doi:10.1088/1742-6596/218/1/012012

Bayesian approach for matching multiple stellar observations

Gyöngyi Kerekes[1], **Tamás Budavári**[2,1], **István Csabai**[1]
[1]Eötvös Loránd University, Department of Physics of Complex Systems, Pázmány P. sétány 1/A, Budapest, 1117, Hungary
[2]Department of Physics and Astronomy, The Johns Hopkins University, 3400 North Charles Street, Baltimore, MD 21218, USA

E-mail: kerekes@complex.elte.hu, budavari@jhu.edu

Abstract. The cross-identification of sources in separate catalogs is one of the most basic tasks in observational astronomy. Recently Budavári & Szalay (2008) formulated the problem in the probability theory, and laid down the statistical foundations of an extendable methodology. An application of the same Bayesian approach to stars is presented that, we know, can measurably move on the sky, and it is shown how to associate their observations. Models are studied on a sample of stars in the Sloan Digital Sky Survey, which allow for an unknown proper motion per object, and their improvements are shown over the simpler static model. The new models and conclusions are directly applicable to the upcoming surveys, whose data sets will be most likely dominated by stars in the region of the Galactic Plane.

1. Introduction

Today the most important astronomical studies use catalog merging as a basic step, so that measurements at different wavelengths by potentially separate instruments and telescopes can be combined. Hence, the cross-identification of sources is becoming an outstanding issue. The case of stars is especially a hard problem. The ones that move between observations are difficult to merge into multicolor objects even within a single survey. Today this poses a significant challenge for the next-generation photometric surveys, such as PanSTARRS and LSST, that plan to cover the Galactic Plane.

In the recent work of (Budavári & Szalay, 2008) a general probabilistic formalism was introduced that is extendable to arbitrarily complex models. The Bayesian hypothesis testing clearly separates the contributions of different types of measurements, e.g., the position on the sky or the colors of the sources, yet, naturally combines them into a coherent method. In this paper, we go beyond the simple case of stationary objects, and study the cross-identification of point sources that move on the sky. Most importantly we focus on stars that, we know, can be significantly offset between the epochs of observations. Although we only have loose constraints on their proper motions in general, this prior knowledge is enough to revise our static models, and work out the Bayesian evidence of the matches.

2. The Bayes Factor

The positional accuracy is characterized by a probability density function (hereafter PDF) on the celestial sphere. In a given model M, this $p(\boldsymbol{x}|\boldsymbol{r}, M)$ function tells us where to expect $\boldsymbol{x}$

detections of an object that is at its true location $\boldsymbol{r}$. We use 3-dimensional unit vectors for the positions on the sky, e.g., the aforementioned $\boldsymbol{x}$ and $\boldsymbol{r}$ quantities.

For hypothesis testing Bayes-factor is used. Let H denote the hypothesis that assumes that all measured positions, D are observations of the same object, and K denote its complement, i.e., any one or more of the detections might belong to a separate object. By definition, the Bayes factor is the ratio of the likelihoods of two hypotheses we wish to compare

$$B(H,K|D) = \frac{p(D|H)}{p(D|K)} \tag{1}$$

that are calculated as integrals of their parameter spaces. If we assume that there is a single object behind the observations, we can integrate over its unknown proper motion and position to calculate

$$p(D|H) = \int\!d\boldsymbol{r}\int\!d\boldsymbol{\mu}\, p(\boldsymbol{r},\boldsymbol{\mu}|H) \prod_{i=1}^{n} p_i(\boldsymbol{x}_i|\Delta t_i, \boldsymbol{r}, \boldsymbol{\mu}, H) \tag{2}$$

where the joint likelihood of H given the data is written as the product of the independent components and $p(\boldsymbol{r},\boldsymbol{\mu})$ is the prior on the parameters, which is the subject of the following section. In the complement hypothesis the integral separates into the product of

$$p(D|K) = \prod_{i=1}^{n} \int\!d\boldsymbol{r}_i\int\!d\boldsymbol{\mu}_i\, p(\boldsymbol{r}_i,\boldsymbol{\mu}_i|K)\, p_i(\boldsymbol{x}_i|\Delta t_i, \boldsymbol{r}_i, \boldsymbol{\mu}_i, K) \tag{3}$$

3. Prior for stars

Based on the basic properties of conditional densities, we can write the prior as the product

$$p(\boldsymbol{r},\boldsymbol{\mu}|H) = p(\boldsymbol{r}|H)\, p(\boldsymbol{\mu}|\boldsymbol{r},H) \tag{4}$$

where the first term is the prior on the position, and the second term describes the possible proper motions as a function of location and optionally other properties. To derive the $p(\boldsymbol{\mu}|\boldsymbol{r},H)$ prior, we choose to study the ensemble statistics of stars instead of approaching the problem with an analytic model, as the formulas are difficult to derive and the analytic approximations might miss subtle details of the relation.

We study the properties of stars in the Sloan Digital Sky Survey catalog archive that also contains accurate proper motion measurements from the recalibrated United States Naval Observatory (USNO) B1.0 Catalog (Munn et al., 2004). For the analysis, we pick stars from the Stripe 82 dataset where multiple observations are available in 300 square degrees of a narrow declination range between $\pm1.26°$ (Adelman-McCarthy et al., 2008). After rejecting saturated and faint sources, the number of stars is around 100,000. This size does not allow for a high-resolution determination of the prior, hence we analyze additional simulation data.

The current state-of-the-art Besançon models are used (Robin et al., 2003) that match the SDSS distributions well. Assuming four different stellar populations in the Milky Way, using Poisson equation and collisionless Boltzmann equation with some observed parameters, (and fitting parameters to the dynamical rotation curve) they compute the number of stars of a given age, type, effective temperature and absolute magnitude, at any place in the Galaxy. $700,000$ stars were generated with Besançon models using large-field equatorial coordinates.

We separate the dependence of the prior on the different proper motion components. Since the Stripe 82 dataset in SDSS contains sources only in a narrow declination range, $-1.26° < \delta < +1.26°$, we can safely neglect the dependence on declination in Equation (4) for the purpose of this study, thus

$$p(\mu_\alpha,\mu_\delta|\boldsymbol{r}) \approx p(\mu_\delta|\alpha)\, p(\mu_\alpha|\mu_\delta,\alpha) \tag{5}$$

Journal of Physics: Conference Series **218** (2010) 012012 doi:10.1088/1742-6596/218/1/012012

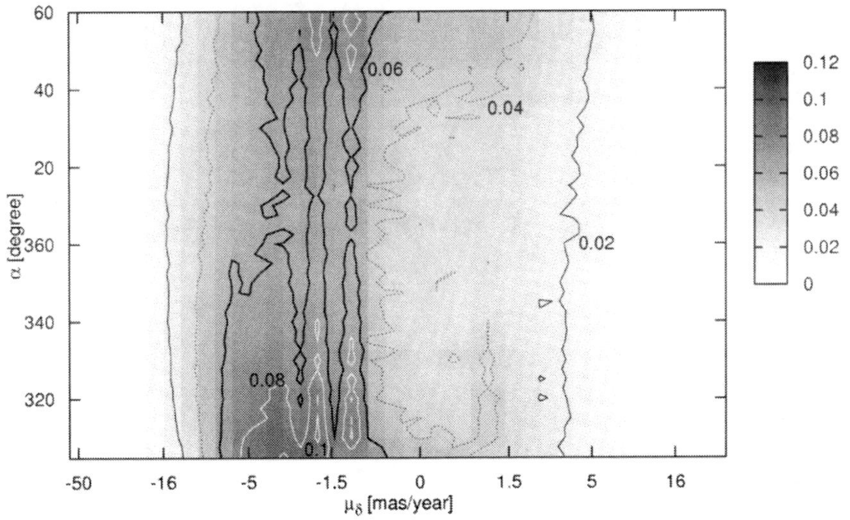

Figure 1. Illustration of the first term of prior (5) on the footprint of SDSS Stripe 82. As one moves into the direction of galactic plane (upwards and downwards on the figure), the probability distribution gets narrower since the velocity dispersion of stars decreases.

To achieve a better signal-to-noise behaviour across the entire parameter space, we do not use a uniform grid but varying bin sizes that follow the asinh() in both parameters. This way one can have higher resolution bins where more data are available and wider bins in the tail. With this transformation $p(\mu_\delta|\alpha)$ becomes an asymmetric shape roughly centered on -0.8 mas/yr. Figure 1 shows the detailed shape of the prior. The second term of Equation (5) is constructed similarly but in even higher dimensions. It is difficult to visualize a 4-dimensional PDF, hence, in Figure 2, we plot slices of the prior at various α values. To evaluate the density for a given position and proper motion, we use linear interpolation on these grids.

4. Results

As the Stripe 82 was observed repeatedly from 1998 to 2005, we can obtain multiepoch observations to test our method. We choose randomly a dozen of stars with different proper motions observed at different epochs. For each star we get on the average 20 epochs, from which many were observed with small time shifts. The longest time intervals between the epochs is on the average 6.5 years, we divide it to 3 roughly equal parts. Thus we get 4 observations of each star with much the same time intervals between them. For the test stars the ObjID marked in Table 1 is the ObjID of `PhotoPrimary` table in SDSS DR6.

In order to calculate the integrals in (2) and (3), we use Monte-Carlo method. The program generates uniformly distributed positions (normal vectors on a sphere) and velocity vectors in a reasonable range. In theory one should integrate for the whole celestial sphere and to infinity in proper motion, hence one can bound all the relevant parameters easily to reduce the computational need. The calculated Bayes-factors are plotted on Figure 3. Applying the means of probability theory, we also calculate the conditional expectations of proper motions for the test stars in case of 4 epochs. On the Figure 4 one can see that using only the observed positions and the information from priors the calculated and the reference values are very close to each other (inside the margin of error).

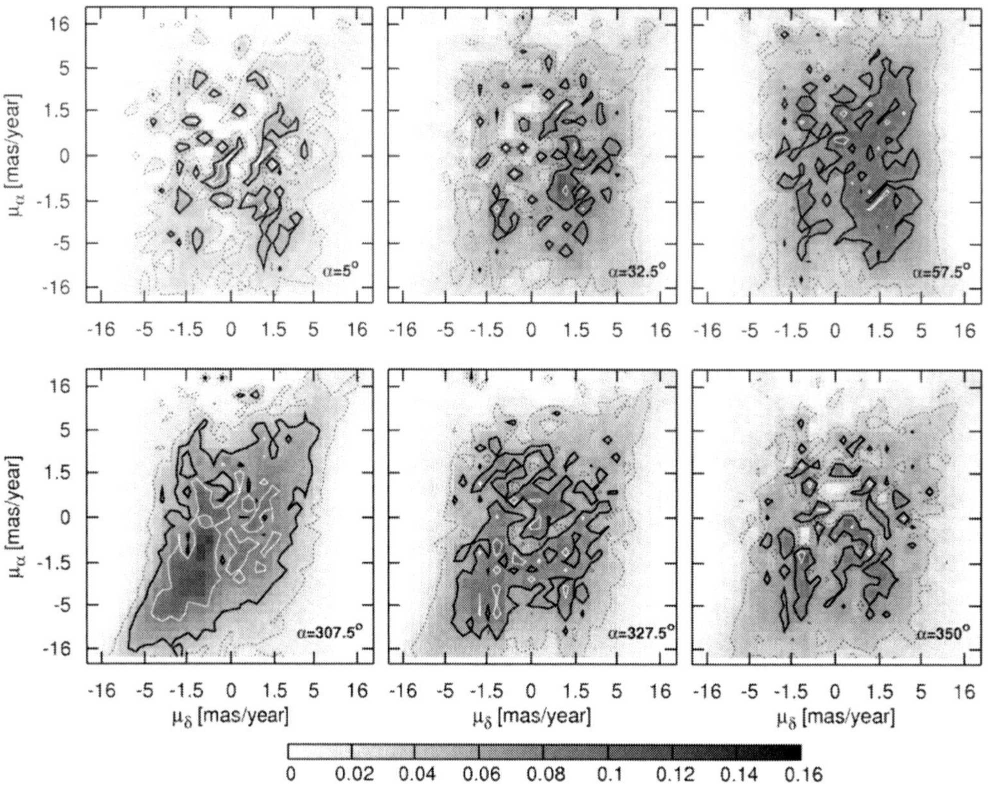

Figure 2. Illustration of the second term of prior (5). The prior is a 4 dimensional PDF, cuts are shown at different RAs. The same effect can be detected as on the Figure 1 - in the cases of $RA = 57.5°$ and $RA = 307.5°$ the distribution is a bit sharper than in the other two cases.

Table 1. Weight of Evidence for Models with no PM and with PM of test stars in case of 2, 3 and observations as a function of USNO proper motions

ObjID	μ [mas/yr]	2 observations static	2 observations motion	3 observations static	3 observations motion	4 observations static	4 observations motion
587731186192679018	2	12.60	12.58	25.32	25.31	38.04	37.97
587730847423987981	8	12.60	12.59	25.26	25.22	37.91	37.86
587730847429427304	13	12.59	12.59	25.25	25.21	38.11	38.08
587731173305876571	18.6	12.59	12.59	25.15	25.12	37.99	37.98
587731186187763779	19	12.55	12.56	25.20	25.23	37.97	37.88
587731173305614418	40	12.59	12.59	25.21	25.22	37.47	37.64
588015509268725910	98	11.93	11.90	22.51	22.86	31.10	34.56
588015509286813878	143	11.64	11.70	19.62	22.20	25.92	34.41
588015509268201645	163	11.44	11.56	18.09	21.84	21.09	34.22
588015509271805995	196	11.32	11.38	10.67	21.48	6.46	33.88
588015509279342731	255	10.02	10.08	7.54	20.01	-5.76	32.60
588015509273378938	257	9.41	9.82	5.44	20.44	-5.48	32.49
587730847426740272	300	9.43	9.63	10.23	20.04	-1.85	31.60
588015509283930154	555	0.40	4.12	-24.19	4.14	-129.09	-18.95

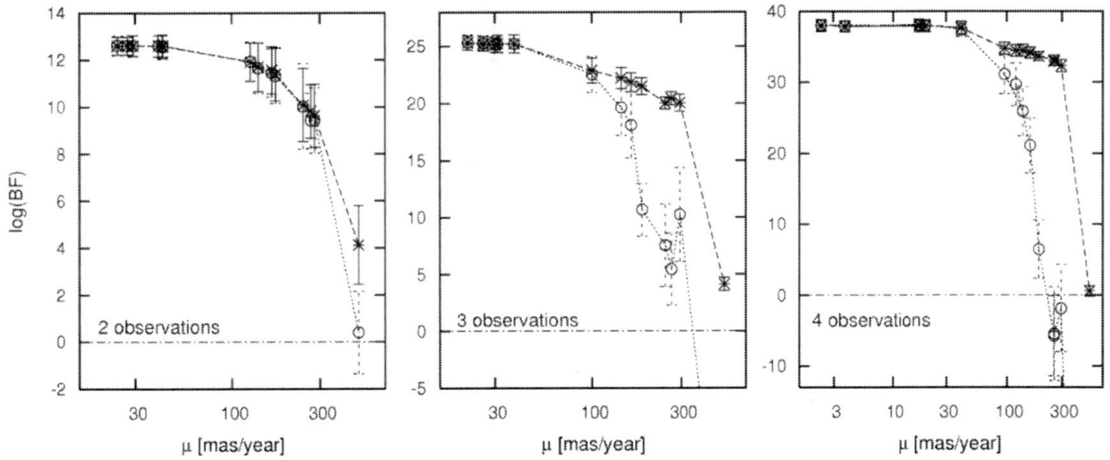

Figure 3. Illustration of the weight of evidence as a function of proper motion in case of 2, 3 and 4 observations. Proper motion is shown on a logarithmic scale. Open circles represent the static model demonstrated in (Budavári & Szalay, 2008). The crosses show the values calculated with the empirical prior introduced in this study. For reference, the W = 0 horizontal line shows the theoretical limit above which the observations support the hypothesis of the match. The error shown on the figure comes from the uncertainty in position measurements.

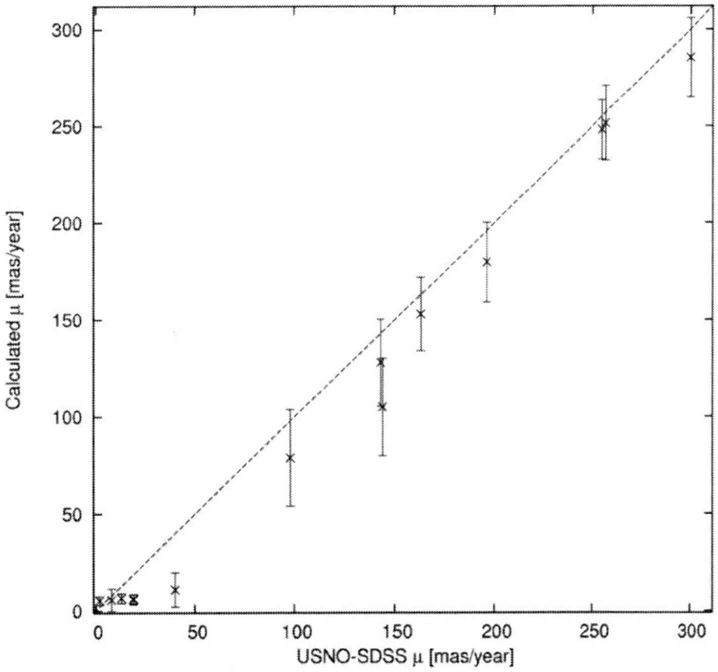

Figure 4. Proper motions calculated for the test stars using the formula of conditional expected value: $E(\boldsymbol{\mu}) = \int d\boldsymbol{r} \int d\boldsymbol{\mu}\, p(\boldsymbol{r}, \boldsymbol{\mu}|H) \prod_{i=1}^{n} p_i(\boldsymbol{x}_i|\Delta t_i, \boldsymbol{r}, \boldsymbol{\mu}, H)$

5. Conclusions

In the models presented here we intended to show that the application of more realistic prior distributions within the Bayesian framework can improve cross-identification of moving astronomical objects. In accord with our expectations, we found that using an empirical prior of the proper motion would assign larger observational evidence to the match. The dependence of the quality of these cross-identifications was studied as a function of separation in time (and space) as well as using multi-epoch observations. The SDSS Stripe 82 sample provided a good test set with 2–4 detections at different times with a few years in between. We found that, even though the 2-epoch data sets benefit from the proper motion model, and the 3-4-epoch observations essentially recover the right associations even for fast-moving stars.

An even more complicated model can include magnitude or color dependence which should improve weight of evidence. Prior could and should be different for every source based on the brightness and color: $p(\mu_\alpha, \mu_\delta|\alpha, C) = p(\mu_\delta|\alpha)\, p(\mu_\alpha|\mu_\delta, C)$ The method can be applied in the following way: first finding objects (galaxies, quasars, etc.) using the static model and for remainders (supposed to be stars) the above-mentioned iterations can be run.

Acknowledgements

The authors would like to acknowledge the use of the online tools of the Besançon collaboration to obtain simulated stars for this study and thank Rosemary Wyse for her invaluable insights and help with stellar model of the Galaxy. G.K. and I.C. acknowledge support from NKTH: Polányi. T.B. acknowledges support from the Gordon and Betty Moore Foundation via GBMF 554.

References

Adelman-McCarthy J K et al. 2008 *ApJS* **175** 297-313

Budavári T, Malik T, Szalay A S, Thakar A R and Gray J 2003 *ASP Conf. Ser.* **Vol. 295** Astronomical Data Analysis Software and Systems XII, eds. H E Payne, R I Jedrzejewski and R N Hook (San Francisco: ASP), p.31

Budavári T et al. 2004 *ASP Conf. Ser.* **Vol. 314** Astronomical Data Analysis Software and Systems XIII, eds. F Ochsenbein, M Allen and D Egret (San Francisco: ASP) p.177

Budavári T and Szalay A S 2008 *ApJ* **679** 301

Fisher R, 1953, Proceedings of the Royal Society of London, Series A, *Mathematical and Physical Sciences* **Vol. 217** No. 1130., pp.295–305

Munn et al. 2004 *AJ* **127** 3034

Robin A C, Reyl C, Derriere S and Picaud S 2003 *A&A* **409** 523

York D G et al. 2000 *AJ* **120** 1579

High resolution studies of the IDV quasar J1128+592

Krisztina Éva Gabányi[1,2], Nicola Marchili[3], Thomas Krichbaum[3], Lars Fuhrmann[3], Xiang Liu[4] and Hua-Gang Song[4]

[1] FOMI Satellite Geodetic Observatory, Budapest, P.O.Box 585 1592 Hungary, [2] Hungarian Academy of Sciences, Research Group for Physical Geodesy and Geodynamics, Budapest, Hungary, [3] Max-Planck-Institute für Radioastronomie, Auf dem Hügel 69, Bonn 53121 Germany, [4] Urumqi Observatory, National Astronomical Observatories, Chinese Academy of Sciences, Urumqi 830011, PR China

E-mail: gabanyik@sgo.fomi.hu

Abstract. J1128+592 is a compact radio quasar and, as ~ 30 % of flat spectrum radio sources, shows fast variations in the cm-wavelength regime on time-scales of an hour to few days. Such phenomenon is often explained by scattering of radio waves in the turbulent, ionized interstellar matter of the Milky Way. Regular monitoring observations of J1128+592 performed by the Effelsberg 100-meter and the Urumqi 25-meter radio telescope during the last four years confirmed this, since a seasonal cycle of the variability time-scale was revealed. This effect is one of the most convincing argument in favor of the interstellar scattering induced fast radio variations in blazars. Multi-epoch, multi-frequency observations made by the Very Long Baseline Array (VLBA) together with the annual modulation model enabled us to determine a lower limit of the distance of the screen responsible for the rapid variability. Additionally, it revealed source-intrinsic–change, which affected the variability pattern.

1. Introduction

Intraday Variability (IDV, Witzel et al. 1986) is a common phenomenon in flat spectrum radio sources, with a fraction of 20-40 % showing this effect. The typical time-scales of the variations range from about 20 minutes to a few days. If the variations originated from the source itself, using causality and the common light-travel time arguments, these short time-scales would invoke extremely small source sizes (in range of μas) and consequently very high brightness temperatures. Brightness temperature values in excess of the inverse-Compton limit (10^{12} K, Kellermann and Pauliny-Toth 1969) are usually explained via relativistic beaming. However, in the case of IDV sources, the invoked Doppler factors are much higher than those derived from different observing methods such as kinetic studies of features in the highly relativistic jets of blazars.

The solution to this controversy is the source-extrinsic theory of IDV (e.g., Rickett 2001 and references therein). This explains that the rapid variations are caused during the propagation of radio waves thus, they do not originate from the source itself. The variations are caused by interstellar scintillation (ISS) in the turbulent, ionized interstellar medium of the Milky Way.

One of the most convincing argument in favor of this propagation theory is the so-called "annual modulation" of the variability time-scale. In the source-extrinsic scenario, the characteristic variability time-scale is inversely proportional to the relative velocity between the observer and the scattering medium. The annual modulation of the variability time-scale

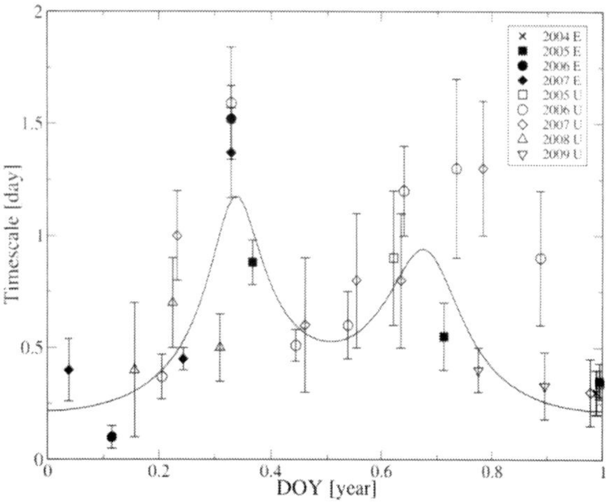

Figure 1. The IDV time-scale of J1128+592 measured at 5 GHz plotted versus day of the year (DOY). Different symbols represent observations performed in different years: star stands for 2004, squares for 2005, circles for 2006, diamonds for 2007, triangles for 2008 and down-triangles for 2009. Filled symbols represent observations performed with the Effelsberg telescope, open symbols for observations with the Urumqi telescope. Line represents the fit of annual modulation model.

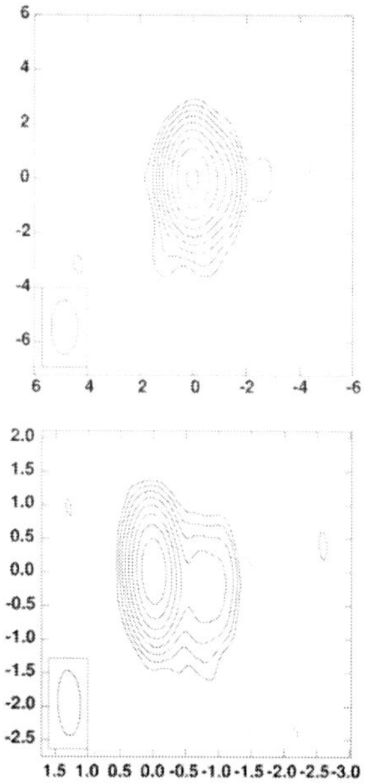

Figure 2. 5 GHz (top) and 15 GHz (bottom) VLBA maps of J1128+592 from epoch 2007.661. The unit of the axes is milli-arcsecond. The peak flux intensities are 299 mJy/beam and 229 mJy/beam at 5 GHz and 15 GHz, respectively. The lowest positive contours are 1 mJy/beam. Contours increase by a factor of two. The beam sizes are shown on the bottom left corners of the images.

reflects the systematic variation in the relative velocity vector between the orbiting Earth (the observer) and the moving scattering screen. Such annual cycle was identified in several IDV sources (e.g., J1819+3845, Dennett-Thrope and de Bruyn 2003, PKS 1622-253 Carter et al. 2009 and references therein).

2. Annual modulation

J1128+592 is an IDV source. During the last four years, the source was regularly observed in more than 30 densely time-sampled flux-density monitoring observations with the Effelsberg 100-meter (MPIfR, Germany) and the Urumqi 25-meter (China) radio telescopes at 5 GHz. (Last observation was carried out in November 2009.) Part of these observations were summarized in previous papers (Gabányi et al. 2007, and references therein). Regarding the observational techniques and data reduction steps we refer to those papers, as well as Fuhrmann et al. (2008) and Marchili (2009).

In most of these observations, J1128+592 showed pronounced variability with peak-to-trough amplitudes exceeding 10 %. The measured characteristic variability time-scales differ significantly from epoch to epoch, ranging from 0.2 to 1.6 days. The different IDV time-scales of J1128+592 measured at different times of the year are indicative of an annual cycle. In the ISS model, the seasonal cycle of the characteristic variability time-scale can be explained as a pure geometrical effect. The scattering material is regarded to be located in a thin plasma screen at some distance from the Earth. The IDV time-scale depends on the relative velocity between the observer and the screen. When the relative velocity is the lowest (the screen and the Earth move approximately parallel), the longest is the IDV time-scale, thus slow variations can be observed. Half year later, when the Earth, due to its orbital motion, moves the opposite direction (thus anti-parallel with the screen), the relative velocity becomes the largest, the observed IDV time-scale is the shortest, thus fast variations can be observed.

In a more general scenario, the scintillation pattern is assumed to be anisotropic. Then the variability time-scale also depends on the ellipticity of the scintillation pattern and the direction in which the relative velocity vector "cuts through" the elliptical scintillation pattern (and not just the absolute value of the velocity vector). The form of the anisotropic annual modulation model (Bignall et al. 2006) which was fitted to the IDV time-scales of J1128+592 is the following: $t(T) = \frac{s\sqrt{r}}{\sqrt{v(T)^2+(r^2-1)(\mathbf{v}(T)\times\mathbf{S})^2}}$, where s is the scintillation length-scale (the product of the angular size of the scintillating source and the screen distance), $\mathbf{v}(T)$ is the relative velocity between the scintillation screen and the observer. As the Earth orbits around the Sun $\mathbf{v}$ varies annually. v denotes the absolute value of $\mathbf{v}$. $\mathbf{S} = (\cos\beta, \sin\beta)$ is the unit vector defining the orientation of the elliptical scintillation pattern. r denotes the axial ratio of the anisotropy. When fitting the model, we derive the velocity of the scattering screen in Right Ascension (v_{RA}) and in Declination (v_δ) direction.

The best fit model to all to observations made so far is shown in Fig. 1. The parameters of the fit: $v_{\mathrm{RA}} = 1 \pm 4\,\mathrm{km/s}$, $v_\delta = -11 \pm 2\,\mathrm{km/s}$, $s = (12 \pm 1) \cdot 10^5\,\mathrm{km}$, $r = 3.0 \pm 0.8$, $\beta = -94° \pm 5°$. (β is measured North through East).

Most of these values are similar to those obtained for other IDV sources showing annual modulation (e.g. J1819+3845, Dennett-Thrope and de Bruyn 2003), however the axial ratios are larger than 4. For example in the case of PKS 1257-326 (Bignall et al. 2006), the axial ratio is ~ 12. Walker et al. (2009) studied whether even higher anisotropy values can give better fits to the measured time-scales in the case of J1819+3845 and PKS 1257-326. They found that the anisotropy can be so large in these two sources that the scintillation pattern can be more effectively described as one dimensional. Therefore, they concluded it would be more likely that the anisotropy is caused by the scattering material rather than the source intrinsic structure. In the case of J1128+592, the axial ratio is not large enough to dismiss the possibility that the anisotropy, observed in the annual modulation, is caused by the source-intrinsic structure.

3. Very Long Baseline Array observations

Between July 2007 and February 2008, six epochs of Very Long Baseline Array (VLBA) observations of J1128+592 were performed. The measurements were carried out at three frequencies: at 5 GHz, 8 GHz and 15 GHz. After correlation at the VLBA correlator in Socorro (NRAO), the data were calibrated and fringe-fitted using the standard procedures within the AIPS (Astronomical Image Processing System) software package. The post-processing of the data included the usual steps of editing, phase- and amplitude self-calibration, and imaging. The Caltech DIFMAP package was used to perform these task and to fit the uv-data. We fitted the source structure in every epoch with circular Gaussian components. More details on the data reduction are given in Gabányi et al. (2009).

These observations revealed that the source has an approximately east-west oriented core-jet structure (Fig. 2). The position angle of the jet component is $\sim -116°$ at 5 GHz, which is remarkably similar to the orientation derived from the annual modulation model of the IDV time-scale. This may indicate that the anisotropy in the scintillation pattern is caused by the source structure rather than the intervening turbulent ISM. However, additional observations are required to clarify unambiguously the roles of the scattering plasma and source structure in the anisotropic scattering. According to Rickett et al. (2002), a negative "overshoot" in the light-curve auto-correlation is indicative of anisotropy caused by the scattering medium.

From the VLBA measurements, we derived an upper limit for the scintillating source size at 5 GHz: 0.2 ± 0.05 mas. Using this, and the scintillation length-scale provided by the annual modulation model, we can derive a lower limit for the distance of the ISM screen responsible for the IDV: ~ 30 pc. This is three times larger than the values derived for the much faster varying J1819+3845 (Denett-Thrope and de Bruyn 2003) and PKS 1257-326 (Bignall et al., 2006). However, it still places the scattering screen within the Local Bubble.

We collected and (re-)analyzed six epochs of archival VLBA observations of J1128+592 made at 5 GHz during 2004. We fitted the source with Gaussian components in order to derive the size of the core. The average core size from 2004 is $\sim 0.07 \pm 0.02$ mas, which is less than half of the value we measured in our VLBA observations in 2007 and 2008.

Our IDV monitoring observations indicated that the variability-strength of the source has declined monotonously since 2006. Additionally, the source underwent a flare-like event, its flux density increased until 2006, then decreased by $\sim 50\,\%$ until late 2007. The strength of variability is usually described by the modulation index (m), which is the ratio of the standard deviation of the flux density to the average flux density. According to the ISS theory of IDV (e.g., Walker 1998), the modulation index goes as $m \sim \theta^{-7/6}$, where θ is the source size. In the case of J1128+592, the modulation index was measured to be around $3-4\,\%$ at the time we had the VLBA observations. Using this equation, we would expect that the source had a modulation index at least 10 % during 2004. Unfortunately, our first single dish observation of the source was made in late December 2004, however the modulation index of J1128+592 was mostly between 7 % and 10 % during 2005. This indicates that the weakening of the strength of IDV can be explained by the combination of the expansion of the VLBI core (or the dominant scintillating component hidden inside) and the decreasing flux density.

Acknowledgments

The National Radio Astronomy Observatory is a facility of the National Science Foundation operated under cooperative agreement by Associated Universities Inc. This paper made use of data obtained with the 100 m Effelsberg radio telescope of the Max-Planck-Institut für Radioastronomie (Bonn, Germany) and the 25 m Urumqi Observatory of the National Astronomical Observatories (NAOC) of the Chinese Academy of Sciences (CAS). This work has benefited from research funding from the Hungarian Scientific Research Fund (OTKA, grant No. K72515).

References

Bignall H E, Macquart J-P, Jauncey D L, Lovell J E J, Tzioumis A K and Kedziora-Chudczer L 2006 *ApJ* **652** 1050-1058
Carter S J B, Ellingsen S P, Macquart J-P and Lovell J E J 2009 *MNRAS* **396** 1222-1230
Dennett-Thrope J and de Bruyn A G 2003 *A&A* **404** 113-132
Fuhrmann L et al. 2008 *A& A* **490** 1019-1037
Gabányi K É, Marchili N, Krichbaum T P, Britzen S, Fuhrmann L, Witzel A, Zensus J A, Müller P, Liu X, Song H G, Han J L and Sun X H 2007 *AN* **328** 863
Gabányi K É, Marchili N, Krichbaum T P, Fuhrmann L, Müller P, Zensus J A, Liu X and Song H G 2009 *A&A* **508** 161-171
Kellermann K I and Pauliny-Toth I I K 1969 *ApJ* **155** L71
Marchili N 2009 Variability studies of a sample of IntraDay Variable sources, PhD Thesis, University of Bonn
Rickett B J, Kedziora-Chudczer L and Jauncey D L 2002 *ApJ* **581** 103-126
Walker M A 1998 *MNRAS* **294** 307
Walker M A, de Bruyn A G and Bignall H E 2009 *MNRAS* **397** 447-452
Witzel A, Heeschen D S, Schalinski C and Krichbaum T 1986 *MitAG Hamburg* **65** 239

5th Workshop of Young Researchers in Astronomy & Astrophysics IOP Publishing
Journal of Physics: Conference Series **218** (2010) 012014 doi:10.1088/1742-6596/218/1/012014

The compact radio structure of the high-redshift blazar J1430+4204 before and after a major outburst

Péter Veres[1,2], **Sándor Frey**[3,5], **Zsolt Paragi**[4,5], **Leonid I. Gurvits**[4]

[1]Dept. of Physics of Complex Systems, Eötvös University, Pázmány P. s. 1/A, H-1117 Budapest, Hungary
[2]Dept. of Physics, Bolyai Military University, POB 15, Budapest, H-1581, Hungary
[3]FÖMI Satellite Geodetic Observatory, POB 585, H-1592 Budapest, Hungary
[4]Joint Institute for VLBI in Europe, POB 2, 7990 AA Dwingeloo, The Netherlands
[5]MTA Research Group for Physical Geodesy and Geodynamics, POB 91, H-1521 Budapest, Hungary

E-mail: `veresp@elte.hu`

Abstract. The high-redshift ($z = 4.72$) blazar J1430+4204 produced an exceptional radio outburst in 2006. We analyzed 15-GHz radio interferometric images obtained with the Very Long Baseline Array (VLBA) before and after the outburst, to search for possible structural changes on milli-arcsecond angular scales and to determine physical parameters of the source.

1. Introduction

Active galactic nuclei (AGNs) are thought to harbor supermassive (up to $\sim 10^{10} M_{\odot}$) black holes. Accretion onto these black holes is responsible for the extreme luminosity of AGNs over the whole electromagnetic spectrum. Part of the infalling matter may be transformed into jets ejected with relativistic speeds. The radio emission in radio-loud AGNs originates from these jets via synchrotron process. If a jet points close to the line of sight toward the observer, its brightness is significantly enhanced. For a review of the unified model of AGNs see Urry & Padovani (1995).

A particular class of AGNs are blazars. They show large variations in brightness from the radio to the gamma-ray regime. They have no emission lines characteristic to other AGNs in the optical spectrum. According to the physical models of AGN (Urry & Padovani 1995), we see at these objects almost exactly in the direction of the jet.

J1430+4204 (B1428+4217) is a blazar with a flat radio spectrum at an extremely high redshift, $z = 4.72$ (Hook & McMahon 1998). Radio flux density monitoring at 15 GHz revealed a significant brightening of J1430+4204, starting in 2004 and reaching its peak in 2006 (Worsley et al. 2006). The object increased its flux density by a factor of 3 in about 4 months (in the source rest frame).

The radio structure of J1430+4204 at the milli-arcsecond (mas) scale is predominantly compact as revealed by Very Long Baseline Interferometry (VLBI) imaging observations (e.g. Paragi et al. 1994, Helmboldt et al. 2007). A weak extension to the bright compact core is also seen in the W-SW direction.

Total flux density outbursts are usually followed by an emergence of a new jet component in the VLBI images of radio AGNs. By observing J1430+4204 after the brightening, we aimed at detecting a new component in a hope to have a zero-epoch point for a later measurement of its apparent proper motion. A study of jet kinematics at such a high redshift would have been interesting since the best-observed sample in the 15-GHz MOJAVE survey (Lister et al. 2007) is restricted to $z < 3.5$.

2. Observations and data processing

We observed J1430+4204 for 8 hours at 15 GHz with the ten 25-m diameter radio telescopes of the NRAO[1] Very Long Baseline Array (VLBA) on 15 Sep 2006. We also found a similar full-polarization experiment (code BY019) in the NRAO data archive. These observations were done on 23 Feb 2005, prior to the brightness peak. We performed standard VLBI calibration, imaging and model-fitting procedures for both data sets using the NRAO Astronomical Image Processing System (AIPS) and the Caltech Difmap program. Our total intensity images are displayed in Fig. 1. The fractional linear polarization of the VLBI core was $\sim$ 1% and 2% at the first and second epoch, respectively.

The source total flux density at 15 GHz was monitored at the Ryle Telescope (Fabian et al. 1999). The light curve is shown in Fig. 2.

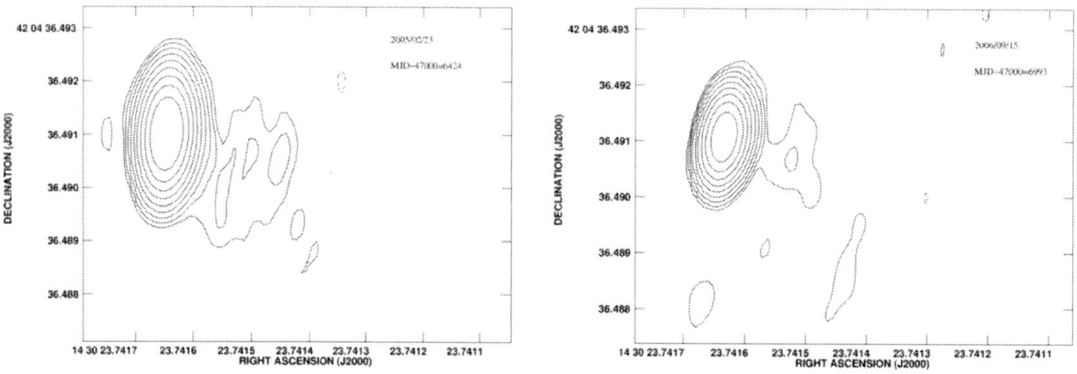

Figure 1. 15-GHz VLBA images of J1430+4204 on 23 Feb 2005 *(left)* and 15 Sep 2006 *(right)*. In the left image, the peak brightness is 200 mJy/beam, the restoring beam is 1.17 mas×0.52mas at the position angle PA=-3.8°. In the right image, the peak brightness is 159 mJy/beam, the restoring beam is 1.22 mas×0.58 mas at PA=-11.95°. In both cases, the lowest contour levels are at 0.3 mJy/beam, the positive contours increase by a factor of 2.

3. Determination of the source parameters

We used three methods for computing the following parameters of J1430+4204: the apparent speed of a possible blob in the jet, $\beta_{app} = \frac{\beta \sin\theta}{1-\beta\cos\theta}$; the Doppler factor, $\delta = \frac{1}{\Gamma(1-\beta\cos\theta)}$; and the Lorentz factor, $\Gamma = \frac{\beta_{app}^2+\delta^2+1}{2\delta^2} = (1-\beta^2)^{-1/2}$. Here $\beta < 1$ is the bulk speed of the material in the jet, expressed in the units of the speed of light c. For the jet angle to the line of sight, $\theta = \arctan\left(\frac{2\beta_{app}}{\beta_{app}^2+\delta^2-1}\right)$, we assumed 3° as found from bulk Comptonization modeling of the observed X-ray spectrum of J1430+4204 (Cellotti et al. 2007) .

[1] The National Radio Astronomy Observatory is a facility of the National Science Foundation operated under cooperative agreement by Associated Universities, Inc.

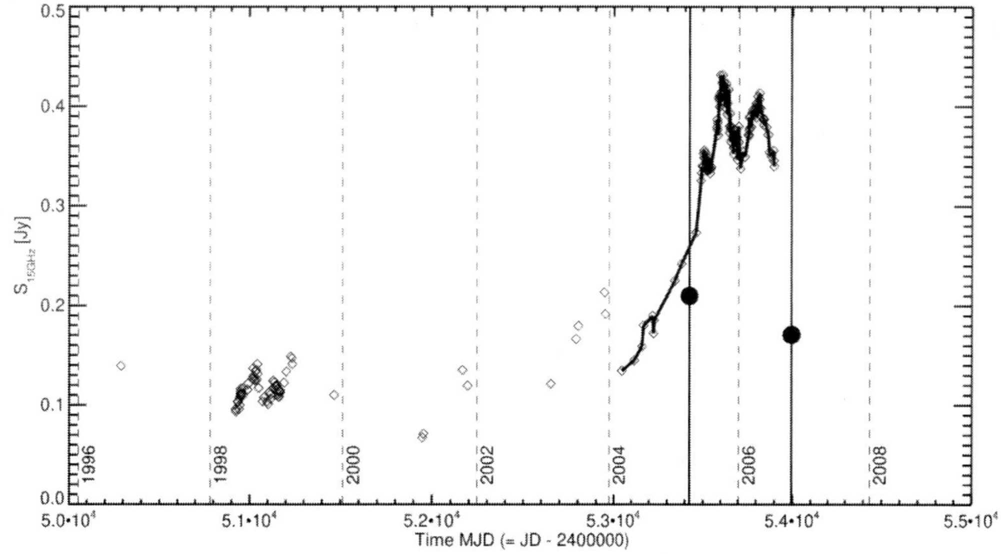

Figure 2. The 15-GHz flux density vs. time for J1430+4204 from the Ryle Telescope monitoring (G. Pooley, priv. comm.). The points are connected in the flaring phase for clarity. Solid vertical lines mark the times of the two VLBA observations, the filled circles correspond to the measured VLBI flux densities. Dashed lines mark calendar years as indicated.

The most important physical parameter characterizing the jet flow is the bulk Lorentz factor (Γ). We are also interested in the apparent tangential velocity of the putative blob in the jet which we translate into proper motion. The distance scale at this redshift is 6.588 pc/mas (assuming a cosmological model with $H_0 = 0.71$ km s^{-1} Mpc^{-1}, $\Omega_\Lambda = 0.73$ and $\Omega_{\rm M} = 0.27$).

3.1. Parameters from the radio variability

We fitted an exponential function to the brightest flare in the flux density curve (Fig. 2) and determined the variability brightness temperature (Hovatta et al. 2009). Furthermore, we took the intrinsic brightness temperature as $T_{\rm b,int} = 5 \times 10^{10}$ K, which assumes equipartition between the particles and the magnetic field in the radio-emitting region (Readhead 1994). We then calculated the so-called variability Lorentz factor and the apparent speed.

We obtained $\Delta S = 412$ mJy for the flare amplitude and $\tau = 458.4^{\rm d}$ for the rise time. These led to the variability brightness temperature $T_{\rm b,var} = 4.1 \times 10^{13}$ K, $\delta_{\rm var} = 9.4$, $\beta_{\rm app,var} = 2.4$, $\Gamma_{\rm var} = 5.1$. The apparent proper motion of the jet component is $\mu = 0.03$ mas/year. This means that during the time between the two VLBA observations ($\Delta T \simeq 570^{\rm d}$) the supposed blob had moved by $\simeq 0.05$ mas. This angular displacement is below the limit which we can possibly detect with the VLBA at this wavelength (~ 0.1 mas).

3.2. Parameters from the brightness temperature measured with VLBI

We fitted circular Gaussian brightness distribution models to the VLBI visibility data at 15 GHz at both epochs. This gave us $T_{\rm b} \simeq 1.2 \times 10^{12}$ K brightness temperatures. It corresponds to a Lorentz factor of $\Gamma \simeq 10$, if we again assume the equipartition value for the intrinsic brightness temperature $T_{\rm b,int}$. This method also predicts a proper motion of $\mu = 0.03$ mas/year.

3.3. Parameters from the inverse Compton process

By assuming that the observed X-ray flux of an AGN is of the inverse Compton origin, we can estimate the Doppler factor (Guijosa & Daly 1994) . For J1430+4204, we took the X-ray data from Celotti et al. (2007) . The formula for the Doppler factor is

$$\delta_{\rm IC} = f(\alpha)(1+z)S_{\rm m} \times \left[\frac{\ln(\nu_{\rm b}/\nu_{\rm op})\nu_{\rm X}^{-\alpha}}{S_{\rm X}\theta_{\rm d}^{6+4\alpha}\nu_{\rm op}^{5+3\alpha}}\right]^{1/(4+2\alpha)} \tag{1}$$

where $f(\alpha) = -0.08\alpha + 0.14$, $S_{\rm m}$ is the radio flux density, $\nu_{\rm b}$ is the synchrotron high-frequency cutoff assumed to be 10^5 GHz. The observed frequency of the radio peak is $\nu_{\rm op} = 15$ GHz, the X-ray flux density is $S_{\rm X} = 3.7 \times 10^{-7}$ Jy taken at $\nu_{\rm X} = 5.2$ keV, $\alpha = -0.4$ is the optically thin spectral index where the $S_\nu \propto \nu^\alpha$ convention is used, and $\theta_{\rm d}$ is the angular diameter of the source in mas. This third method also provided $\mu \simeq 0.03$ mas/year for the jet component proper motion.

4. Discussion and conclusion

The high-redshift blazar J1430+4204 produced an exceptional radio flux density outburst in 2006 (Fig. 2). We imaged the source with the VLBA at 15 GHz after the time of the flux density peak, and also analyzed the archive VLBA data taken during the rise of the total flux density curve. At both epochs, the mas-scale radio structure of the source was similar: a compact core and a weak extension to S-SW (Fig. 1). The core could be fitted with circular Gaussian components with sizes of 0.076 mas and 0.060 mas (full width at half maximum) on 23 Feb 2005 and 15 Sep 2006, respectively. The comparison of the total and VLBI flux densities at the first epoch (Fig. 2) suggests that ~ 50 mJy could be attributed to the radio emission extending to more than a few mas.

Based on our VLBA imaging, we do not detect any new separate jet component to be associated with the outburst. Assuming a small jet angle to the line of sight, we used three different methods to calculate the expected proper motion of such a component. These gave consistently small values of the proper motion. We conclude that our time base and angular resolution were insufficient to distinguish any new blob in the jet. Our estimates for the bulk Lorentz factor ($\Gamma \simeq 5-10$) are comparable with the typical values found for other blazars (e.g. Hovatta et al. 2009).

Acknowledgments

This work was supported by OTKA grants K077795 and K72515. We are grateful to Guy Pooley for sharing the Ryle Telescope data with us.

References

Celotti A, Ghisellini G and Fabian A C 2007 *MNRAS* **375** 417
Fabian A C, Celotti A, Pooley G, et al. 1999 *MNRAS* **308** L6
Guijosa A and Daly R A 1996 *ApJ* **461** 600
Helmboldt J F, Taylor G B, Tremblay S, et al. 2007 *ApJ* **658** 203
Hook I M and McMahon R G 1998 *MNRAS* **294** L7
Hovatta T, Valtaoja E, Tornikoski M and Lahteenmaki A 2009 *A&A* **494** 527
Lister M L, Cohen M H, Homan D C, et al. 2009 *AJ* **138** 1874
Paragi Z, Frey S, Gurvits L I, et al. 1999 *A&A* **344** 51
Readhead A C S 1994 *ApJ* **426** 51
Urry C M and Padovani P 1995 *PASP* **107** 803

Probing the Stellar Populations in the Outskirts of Spiral Galaxies

Judit Bakos[1], **Ignacio Trujillo**[1], **and Michael Pohlen**[2]
[1]Instituto de Astrofísica de Canarias, C/ Vía Lactea S/N, La Laguna, 38205, Spain
[2] School of Physics & Astronomy Cardiff University, CF24 3AA, Wales, The United Kingdom
E-mail: jbakos@iac.es

Abstract. We present our results on the stellar population properties of the outskirts of disk galaxies. In particular, we focus on spiral galaxies with stellar disk truncations in their radial surface brightness profiles. Using SDSS data we show color gradients. We find that the color radial profile exhibits a "U"-shape", showing a minimum at the position of the break of the surface brightness profile. We obtain stellar surface mass density profiles of truncated galaxies as well, these show a peculiar behaviour: they follow very closely an exponential decrement. This suggests the idea that the observed properties of truncated galaxies are not caused by a drop in the mass distribution but by a different stellar population in the outer regions of the disks. Confronting this with current theoretical scenarios we find that this is likely to be a result of secular evolution, in which scenario stars are being formed inside the break and then being motioned outwards which would result in an inverted age gradient corresponding to the observed color profiles. Having maintained this idea, using multiwave-length data from GALEX, SDSS, UKIDSS and SPITZER we showed a dependence of the inner-outer scale-length ratio of truncated disks. Our results suggest that there is an existing general trend of the scale-length ratio: from bluer to redder bands the scale-length ratio decreases, which is in accordance with the stellar disk being dominated by an older stellar population in the outer disk.

keywords: *Galaxies: Evolution* – *Galaxies: Structure* – *Galaxies: Formation* - *Galaxies: Spiral* - *Galaxies: Photometry*

1. Introduction

Multiband observations (Gil de Paz et al., 2005; Pohlen & Trujillo, 2006; Erwin et al. 2008) show evidence of a large number of stars being present in the outer regions of spiral galaxy disks. However, current star formation theories do not support the idea of stars being created in those regions. Because the environment is not dense enough to provide the conditions of star formation, e.g. gas surface mass density is too low ($\leq 10\ M_{\odot}pc^{-2}$, Kennicutt, 1989). This fact creates a so-far not answered question: what is the origin of these stars? We approach this problem by means of investigating the structural and stellar population properties of the (outskirts) spiral galaxy disks.

Early studies of the disks of spiral galaxies (Patterson, 1940; de Vaucouleurs, 1958; Freeman, 1979) showed that this component generally follows an exponential radial surface-brightness profile, with a certain scale-length, usually taken as the characteristic size of the disk. Freeman (1970) pointed out, though, that not all disks follow this simple exponential law. In fact, a repeatedly reported feature of disks for a representative fraction of the spiral galaxies is that of a truncation (Van der Kruit, 1979) of the surface brightness at large radii, typically 2-4 exponential scale-lengths (see e.g. the review by Pohlen et al., 2004).

Several possible break-forming mechanisms have been investigated to explain the truncations. There have been ideas based on maximum angular momentum distribution: Van der Kruit (1987), proposed that angular momentum conservation in a collapsing, uniformly rotating cloud naturally gives rise to disk breaks at roughly 4.5 scale radii. Van den Bosch (2001), suggested that the breaks are due to angular momentum cut-offs of the cooled gas. On the other hand, breaks have also been attributed to a threshold for star formation (SF), due to changes in the gas density (Kennicutt, 1989), or to an absence of equilibrium in the cool Interstellar Medium phase (Elmegreen & Parravano, 1994; Schaye, 2004). Magnetic fields have been also considered (Battaner et al., 2002) as responsible of the truncations. More recent models using collisionless N-body simulations, such as that by Debbatista et al. (2006), demonstrated that the redistribution of angular momentum by spirals during bar formation also produces realistic breaks. In a further elaboration of this idea, Roškar et al. (2008) have performed high resolution simulations of the formation of a galaxy embedded in a dark matter halo. In these models, breaks are the result of the interplay between a radial star formation cut-off and redistribution of stellar mass by secular processes. A natural prediction of these models is that the stellar populations present an age minimum in the break position. This prediction could be probed by exploring the optical color profiles of the galaxies. Another approach is to use fully cosmological simulations of disk galaxy formation. The latest models (Sánchez-Blázquez et al. 2009; Martínez-Serrano et al. 2009) are able to explain all the observational phenomenology and predict the break to be a consequence of the appearance of warp at the truncation radius. This warp will decrease the gas density beyond the break decreasing the star formation and consequently producing a break in the surface brightness profile. A prediction of these models is that there should be an increase in the gas velocity dispersion beyond the break. This is explored in great detail in Trujillo & Bakos (2010, in prep.). Furthermore, exploring multiwave-length data in order to study how the break properties change depending on the observing wavelength can prove to be important in our study of the dominant stellar population inside the disks.

Here we summarize the results of Bakos, Trujillo & Pohlen, 2008 (BTP08, later on) and Bakos & Trujillo, 2010 (in prep).

2. Color Profiles of Local Galaxies

In order to constrain the outer disk formation models, in BTP08 we have explored radial color and stellar surface mass density profiles for a sample of 85 late-type spiral galaxies with available deep (down to $\sim$27 mag/arcsec^2) SDSS g'- and r'-band surface brightness profiles (Pohlen & Trujillo, 2006). About 90% of the light profiles have been classified as broken exponentials, exhibiting either truncations (Type II galaxies) or antitruncations (Type III galaxies). Their associated color profiles show a significantly different behavior. For the truncated galaxies a radial inside-out bluing reaches a minimum of $(g' - r') = 0.47 \pm 0.02$ mag at the position of the break radius, this minimum is followed by a reddening outwards (see middle row in Fig. 1). The antitruncated galaxies reveal a different behavior. At the position of the break radius (obtained from the light profiles) now resides a plateau region of the color profile with a value about $(g' - r') = 0.57 \pm 0.02$.

2.1. Stellar Surface Mass Density Profiles

Using the $(g' - r')$ color it is possible calculate mass-to-light (M/L) ratios along the radius of the disks (Bell et al., 2003). Converting the (M/L) into stellar surface mass density reveals a surprising result. The breaks, well established in the light profiles of the Type II galaxies, are almost gone (in case of several individual galaxies, e.g., NGC5300, the break is completely gone). The mass profiles resembles now those of the pure exponential Type I galaxies (see bottom row in Fig. 1). This result suggests that the origin of the break in Type II galaxies is more likely

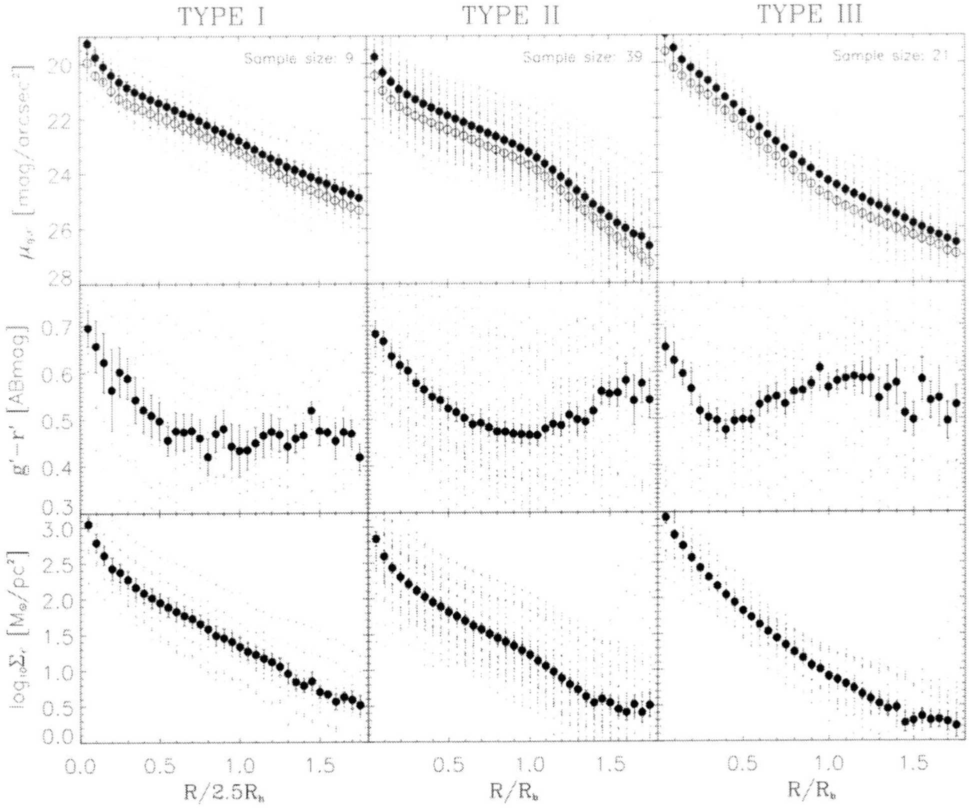

Figure 1. *Upper row*: Averaged, scaled radial surface brightness profiles of 9 Type I (pure exponential profiles), 39 Type II (truncated galaxies) and 21 Type III (antitruncated) galaxies. The filled circles correspond to the r'-band mean surface brightness, the open circles to the mean g'-band data (PT06). The small dots are the individual galaxy profiles in both bands. The surface brightness is corrected for Galactic extinction. — *Middle row*: $(g' - r')$ color gradients. The averaged profile of Type I reaches an asymptotic color value of ~ 0.46 mag being rather constant outwards. Type II profiles have a minimum color of 0.47 ± 0.02 mag at the break position. The mean color profile of Type III has a redder value of about 0.57 ± 0.02 mag at the break. — *Bottom row*: r'-band surface mass density profiles obtained using the color to M/L conversion of Bell et al. (2003), and using *Kroupa-IMF* (Kroupa, 2001). Note how the significance of the break almost disappears for the Type II (truncated galaxies) case.

due to a radial change in the ingredients of the stellar population than being associated to an actual drop in the distribution of mass. The antitruncated galaxies, on the other hand, show clear mass-excess in the outer regions on the stellar mass density profiles, which could have been accumulated from an external (possibly satellite) origin.

There are other structural parameters that can be computed to constrain the different formation scenarios. Among these we have estimated the stellar surface mass density at the break for truncated (Type II) galaxies (13.6±1.6 $M_{\odot}pc^{-2}$) and the same parameter for the antitruncated (Type III) galaxies (9.9±1.3 $M_{\odot}pc^{-2}$). Finally, we have measured that ~15% of the total stellar mass in case of truncated galaxies and ~9% in case of antitruncated galaxies is found beyond the measured break radii in the light profiles.

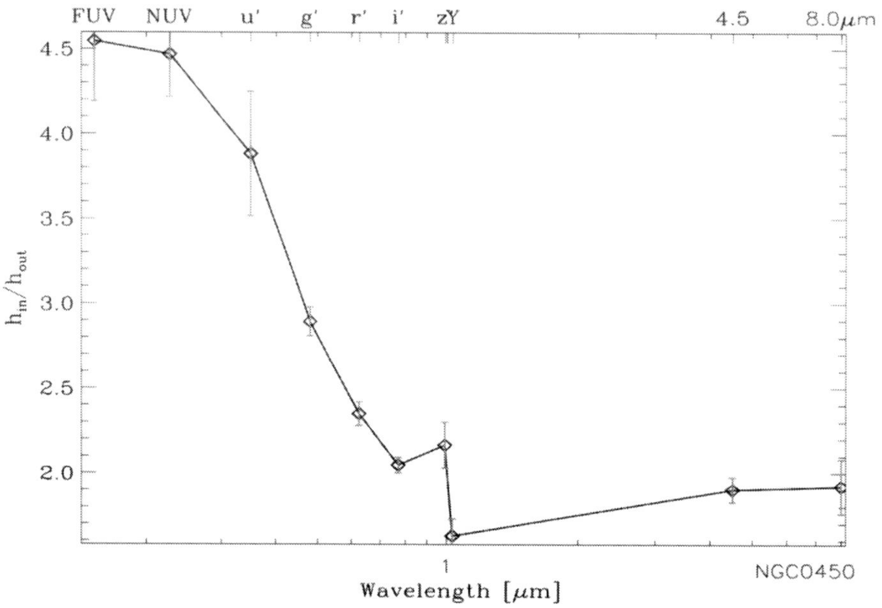

Figure 2. The dependence of the scale-length ratio on wavelength in case of NGC0450. The decrease on the ratio towards redder wavelength is quite dramatic. Errorbars are propagated errors coming from the errors on the fit. We fitted the surface brightness profiles by using robust linefitting method.

3. Multi-wavelength characterization of a truncated disk

To further investigate the origin of the surface brightness break, we have chosen galaxies with full wavelength coverage (from UV to IRAC) from the PT06 sample. Here we show an example of NGC0450. We have obtained the radial surface brightness profiles in all bands using fixed isophotes of the ellipticity and position angle, see Fig. 3. We first masked the stars and other background objects on the galaxy image, the objects were identified by SExtractor (Bertin & Arnouts, 1996). After masking we calculated the moments of the light distribution of the galaxy in r'-band, which is sufficiently smooth enough to derive the correct position angle and ellipticity of the galaxy. The break radius of this galaxy is independent of the wavelength and it is located at $\sim$ 80 arcseconds. (See PT06, and Bakos & Trujillo (2010, in prep).)

We characterize the outer disk by means of calculating the ratio of the inner and outer scale-lengths.We use profiles which are deep enough to obtain reliable fit beyond the break. We find that the scale-length ratio decreases towards redder wavelengths.(See Fig. 2.)

4. Discussion

Our results on the color profiles fit qualitatively with the particular prediction of Roškar et al. (2008), that the youngest stellar population should be found at the break radius, and older (redder) stars must be located beyond that radius. It is not easy to understand how "angular momentum" or "star formation threshold"/"ISM phases" models alone could explain our results. Thus they pose a difficult challenge for these models. However, it will also be necessary to check whether the Roškar models (as well as other available models in the literature like those of Bournaud et al. (2007) and Foyle et al. (2008) are able to reproduce quantitatively the results shown here.

Nevertheless, the fact that the stellar mass density of the break for Type II galaxies with $\sim 13 M_{\odot} pc^{-2}$ is so close to the gas density threshold prediction of $\sim 10 M_{\odot} pc^{-2}$ makes the case for a stellar population origin for the surface brightness break even stronger (with a 100% efficiency of transforming gas to stars).

Our results on the dependence of the inner-outer scale.length ratio on the observing wavelength suggests that indeed this is the case: from bluer to redder bands the scale-length ratio decreases, which is in accordance with the stellar disk being dominated by an older stellar population in the outer disk.

Evidence for the same color phenomenology for Type II galaxies also at high redshift is presented by Azzollini et al. (2008). They have shown that a similar minimum in the color profile can be found, at least, up to z$\sim$1 and that the main source of the scatter of the color profiles is caused by the different stellar mass (in our case absolute magnitude) of the galaxies in their sample. Following these findings, we can conclude that once the absolute magnitude of a galaxy is fixed, the color profiles within a given Type (I, II or III) of galaxy, are strikingly similar.

Combining the results found in Azzollini et al. (2008b) and BTP08 one is tempted to claim that both the existence of the break in Type II galaxies, as well as the shape of their color profiles, are long lived features in the galaxy evolution. Because it would be hard to imagine how the above features could be continuously destroyed and re-created maintaining the same properties over the last $\sim$8 Gyr.

References

Azzollini R, Trujillo I, and Beckman J E 2008a, *ApJ* **684** 1026
Azzollini R, Trujillo I, and Beckman J E 2008b, *ApJ* **679** L69
Bakos J, Trujillo I, and Pohlen M 2008, *ApJ*, **683**, L103 (BTP08)
Bakos J and Trujillo I 2010, in prep
Battaner E, Florido E, and Jiménez-Vicente J 2002, *A&A* **388** 213
Bell E F, McIntosh D H, Katz N, Weinberg M D 2003, *ApJS* **149** 289
Bertin E and Arnouts S 1996, *A&AS* **117** 39
Bournaud F, Elmegreen B G, and Elmegreen D M 2007, *ApJ* **670** 237
Debattista V P, Mayer L, Carollo C M, Moore B, Wadsley J, and Quinn T 2006, *ApJ* **645** 209
de Vaucouleurs G 1958, *ApJ* **128** 465
Elmegreen B G and Parravano A 1994, *ApJ* **435** L121
Erwin P, Pohlen M, and Beckman J E 2008, *AJ* **135** 20
Foyle K, Courteau S, and Thacker R J 2008, *MNRAS* **386** 1821
Freeman K C 1970, *ApJ* **610** 811
Gil de Paz A, Madore B F, Boissier S, Swaters R, Popescu C C, Tuffs R J, Sheth K, Kennicutt R C Jr., Bianchi L, Thilker D, and Martin D C 2005, *ApJ* **627** L29
Kennicutt R C 1989, *ApJ* **344** 685
Kroupa P 2001, *MNRAS* **322** 231
Martínez-Serrano F J, Serna A, Doménech-Moral M, and Domínguez-Tenreiro R, 2009, *ApJ* **705** L133
Patterson F S 1940, *Harvard Coll. Obs. Bul.* 914 9
Pérez I 2004, *A&A* **427** L17
Pohlen M, Beckman J E, Hüttemeister S, Knapen J H, Erwin P, and Dettmar R-J 2004, *Penetrating Bars through Masks of Cosmic Dust: The Hubble Tuning Fork Strikes a New Note*, ed. D. L. Block, I. Puerari, K. C. Freeman, R. Groess, and E. K. Block (Dordrecht: Springer) 731
Pohlen M, and Trujillo I 2006, *A&A* **454** 759
Roškar R, Debattista V P, Stinson G S, Quinn T R, Kaufmann T, Wadsley J 2008, *ApJ* **675** L65
Sánchez-Blázquez P, Courty S, Gibson B K, and Brook C B 2009, *MNRAS* **398** 591
Schaye J 2004, *ApJ* **609** 667
Trujillo I and Bakos J 2010, in prep
van den Bosch F C 2001, *MNRAS* **327** 1334
van der Kruit P C 1979, *A&AS* **38** 15
van der Kruit P C 1987, *A&A* **173** 59

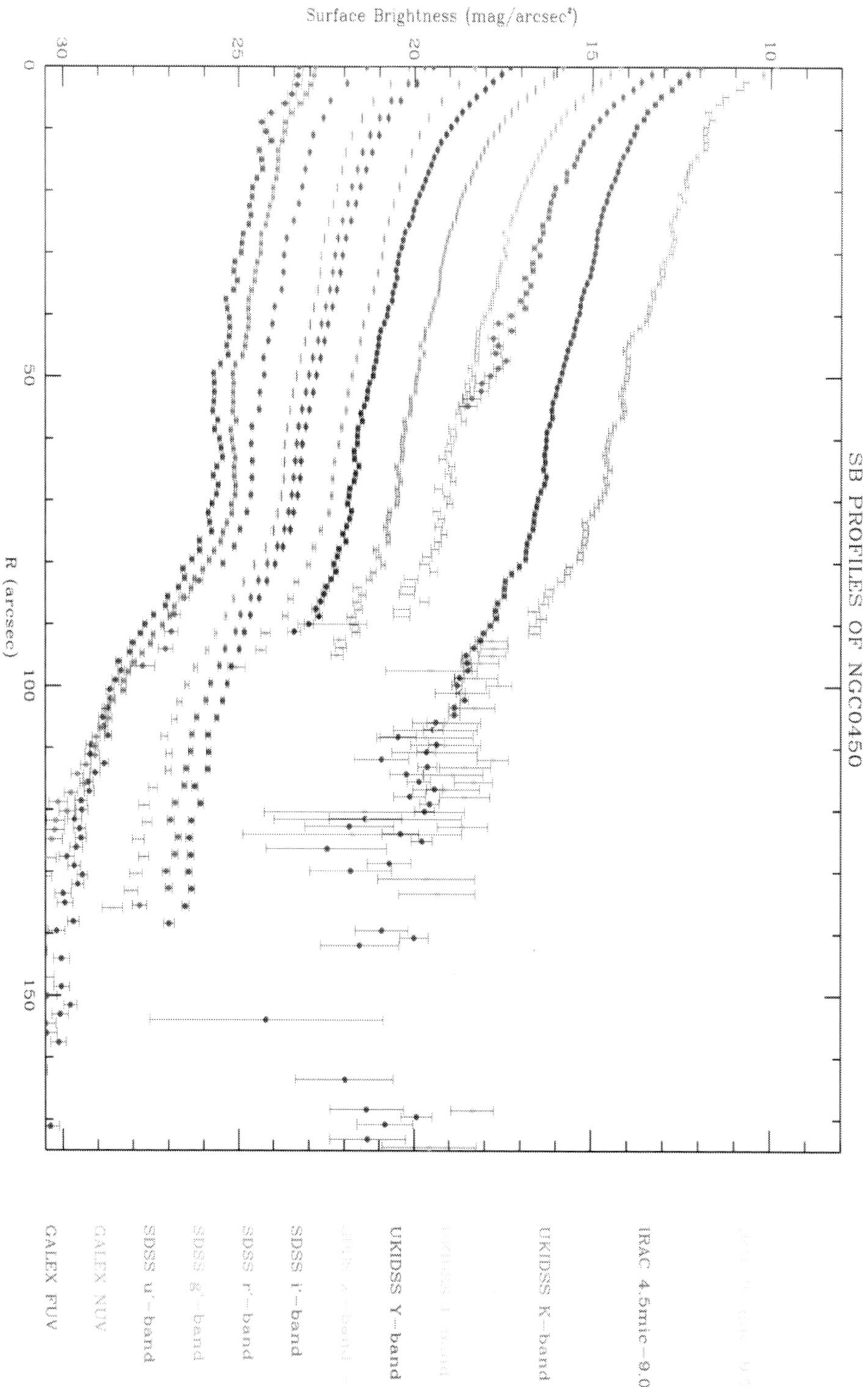

Figure 3. Radial surface brightness profiles of NGC0450 obtained from UV to IRAC. The profiles are shown down to the surface brightness limit of a given filter.

5th Workshop of Young Researchers in Astronomy & Astrophysics IOP Publishing
Journal of Physics: Conference Series **218** (2010) 012016 doi:10.1088/1742-6596/218/1/012016

Energy use, entropy and Extra-terrestrial Civilizations

Zsolt Hetesi[1]
Teacher of Cosmology, Eötvös University Department of Astronomy
Budapest H-1518 P.O. Box: 32

E-mail: zs.hetesi@astro.elte.hu

Abstract.
The possible number of extra-terrestrial civilizations is estimated by the Drake-equation. Many articles pointed out that there are missing factors and over-estimations in the original equation. In this article we will point out that assuming some axioms there might be several limits for a technical civilization. The key role of the energy use and the problem of the centres and periphery strongly influence the value of the L lifetime of a civilization. Our development have several edifications of the investigations of the growth of an alien civilization.

1. Introduction

One of the most puzzling questions of mankind is whether there are other civilizations in the Cosmos and the many research activities that address this subject are generically that called Search for Extraterrestrial Intelligence (SETI).

The number of actual civilizations in our Galaxy can be estimated (we will not interested in extragalactic scales in accordance with the original goal of Drake) by the Drake Equation, which was outlined by Dr Frank Drake as a focus for the Green Bank Conference in 1961. The equation contains the most important factors regarding the necessary conditions of intelligent life and is normally expressed as:

$$N = R_* n_p f_{pl} f_l f_{civ} L \tag{1}$$

where N is the number of intelligent civilizations in our Galaxy, R_* means the rate of star formation, n_p is the number of planets in the habitable zone around a star, f_{pl} is the fraction of habitable planets, f_l is the fraction of life-bearing planets, f_{civ} is the fraction of civilized ones, and finally L is the average lifetime of an intelligent civilization (Wallenhorst 1981, Musso 2001).

We pointed out earlier that some of these factors over-estimate N and there are missing factors in (1) (Regály and Hetesi 2006). If we assume that there exist some axioms valid on every extra-terrestrial (ET) civilizations we are able to give a new estimation of the L_{com} factor. This factor was introduced in Regály and Hetesi (2006) and means the communicative lifetime of an alien civilization. The majority of the matter of a rocky planet is from the dust of a supernova explosion therefore likely contains elements useful for nuclear fission. Lithium which

[1] To whom any correspondence should be addressed.

is also convenient for energy production via nuclear fusion was synthesized in the primordial nucleon synthesis. Its abundance is matter of disputes (Suzuki et al. 2000) and may changed by giant stars (Sackmann and Boothroyd 1995). We assume here that a rocky planet has the same amount of lithium source as the Earth (and we allow only one order of magnitude step in each direction).

If the Darwinian evolution is correct, and the role of the Maximum Power Principle (MPP) is valid in every ET ecosystems we must assume

- An exponential growth in technology, population, energy usage while abundant energy resources are present.
- Shorter life intervals for these civilizations.

In the third section we investigate our axioms on the role of the evolution and the MPP. After this we add together the time intervals of different energy sources as main sources in an ET society. Finally we investigate the role of the centres and peripheries in a society which has no more available abundant energy sources. It contains some consequences for the communicative lifetime of a civilization like this.

2. Axioms and suppositions

In this section we discuss how many axioms we need to imagine a technical civilization on a distant rocky planet.

As we know the Darwinian evolution describes the development of the species on the Earth but it may very possibly regulate the evolution of organic forms everywhere where life once emerged. Therefore its universal validity is our first axiomatic assumption.

The second assumption is the presence and validity of the MPP. It was introduced and detailed in Odum (1983) and Kaila and Annila (2008) and states that species which use the available energy with greater speed can evolve faster. In other words the most important attribute of the evolution in consideration of energy use that evolution follows the steepest descents in the energy use. If a technical civilization was able to emerge from the natural environment it uses more energy than which is naturally available. This type of a society can cross the limits which come from the operation of the natural ecosystem. Therefore this society can grow until the energy sources will be depleted. I.e. our second axiomatic assumption is the presence and operation of the MPP.

Finally we assume that the evolution is a long process and contains many dead ends. Hence there are many (possibly carbon based) remnants under the surface which accumulate energy.

3. Energy use and technical civilizations

The continuous growth of a society causes growing energy demand. If the axioms which were detailed are correct, we can expound the connection between the development and energy use. The second law of thermodynamics can be interpreted alternatively: processes decrease the work-performing capability of the energy. I.e. the MPP guarantees that the

$$\frac{\partial S}{\partial t} = max \tag{2}$$

partial derivative will have its maximum value. If the equalization is finished there will be no more thermodynamic free energy therefore the system is in its lower available state. Hence the economic (or social, technical etc.) growth is purely the continuation of the evolution of species in human circumstances and the consequence of the MPP because mankinds efforts to grow as fast as possible are in accordance with the MPP and equation (2).

Therefore the total amount of an accumulated energy source (signed by R) lasts for

$$T_{use} = \frac{1}{k} ln \left(\frac{r_0}{R} + 1 \right) \tag{3}$$

The k means the fractional growth in the consumption per year, r_0 is the initial consumption of the investigated source.

Table 1. Different energy sources and their usage times

Type	Assumed amount (Gigatons)	Usage time (years)
Hydrocarbons	$10^5 - 10^{10}$	$10^2 - 10^4$
Fissile materials	$10^{-2} - 10$	$10^2 - 10^3$
Lithium 6	$10^{-3} - 10^2$	$10 - 10^3$
Lithium total	$10^{-2} - 10^3$	$10^2 - 10^4$
Materials from the planetary system	$10^{10} - 10^{58}$	$10^4 - 10^9$

We assumed that the transitions from a depleting source to a richer one happen when the depletion just have started, i.e. the growth is continuous. The amount of these sources allows a technical civilization to exist for $10^4 - 10^9$ years order but there will be a (recently unknown) distribution according to the age at the collapse. There might civilizations which did not collapse and presently are active.

4. New estimation of L_{com}

The L_{com} means the time interval in which we can accept signals from a non-silent technologically developed ET civilization. It was originally assumed that whenever a civilization enters the realm of communicative technologies it will remain there for indefinite time. Therefore L_{com} is estimated to be in the orders of billion years.

However if our chain of thought is correct, there were civilizations which collapsed due to a crisis caused by energy shortages and/or ecosystem collapse. These events might be more common in the Universe as the MPP is valid and makes societies to grow and use more energy and raw materials. If a societys growth is led by apparently abundant resources the control of nature ceases over it and the society does not recognize feedbacks from the nature. Therefore the short term regulators do not have any effect on the growth hence this growth exceeds the sustainable level. When the apparently abundant source is gone society collapses.

According to these results L_{com} might be with 0-6 orders of magnitude smaller. In order to avoid giving irresponsible estimations we have to consider the following:

- There are possibly other abundant energy sources in a planetary system. If a society is able to use the lithium based fusion for 104 years with growth rate of 5%, it will be 714 times bigger energy consumer at the end. It means the colonization of the near space.
- It is possible to construct a sphere around the central star and use all of its energy. In this case there will be a sphere with IR-radiaton in the centre of this planetary system.
- If there is failure during an energy source change due to technological gaps the collapse may occur.
- Not only the amount of energy sources counts but their flow is also important. If there is a bottleneck effect on this field for a growing civilization it may collapse.

Journal of Physics: Conference Series **218** (2010) 012016 doi:10.1088/1742-6596/218/1/012016

We do not know any of these civilizations but it is not reasonable to suppose that all of them have been able to solve the problem of energy source transitions. Therefore it is very likely that Lcom might be with 0-6 orders smaller.

5. Collapse due to energy shortages. Distribution of goods

If an ET societys growth is in danger due to an energy crisis there is a new possibility to avoid the sudden collapse and the centres of these civilizations will be able to continue growth (Homer-Dixon 2006). The distribution of goods in a growing society where MPP is valid follows a power law graph.

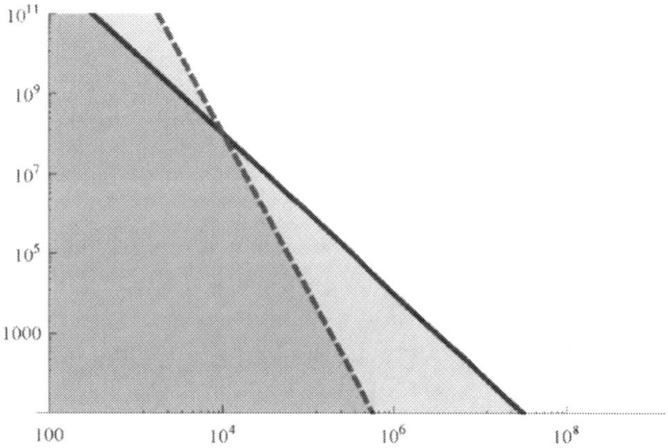

Figure 1. Power law distribution of goods in a society similar to ours. If we consider the group of riches with a step of one magnitude more richness there will be a step downward in the magnitude of the number of the persons who owe that richness. If the centre takes energy from the periphery it can continue its growth, i.e. the steepness of the graph increases (dashed line). Meanwhile the periphery sinks into poverty. The amount of the goods (the integral of the two functions) remains the same or slowly decreases. If there is severe tension accumulated on the periphery collapse occurs.

6. Conclusion

We introduced some axiomatic assumptions in order to investigate the role of entropy and MPP in the biological evolution and societal development. We pointed out that the growth of a technical civilization is strongly connected with the Maximum Power Principle and the growth is only a natural consequence of this principle. L_{com} may be smaller with several orders of magnitude because of the limited energy sources and/or flows. These two phenomena may cause the end of a civilization. More precise estimations are needed to calculating the effect of these phenomena on L_{com}.

References

Homer-Dixon T 2006 *The Upside of Down* (Island Press)
Kaila V R I and Annila A 2008 *Proc. R. Soc. A* **464** 3055
Musso P *On the last terms of Drake equation: the problem of energy sources and the "rare Earth hypothesis"* In First European Workshop on Exo- Astro-Biology, Frascan 21-23 May 2001, ESA-SP 496, August 2001
Odum H T 1983 *Systems Ecology: An Introduction* (New York: Wiley)
Regály Zs and Hetesi Zs 2006 A New Interpretation of Drake-Equation *JBIS* **59** 11
Sackmann I J and Boothroyd A I 1995 Lithium Creation In Giant Stars *MmSAI* **66** 403
Suzuki T K, Yoshii Y and Beers T C 2000 Primordial Lithium Abundance as a Stringent Constraint on the Baryonic Content of the Universe *ApJ* **540** 99
Wallenhorst S G 1981 *QJRAS* **22** 380-87

Stability of the orbit of a third body in binary asteroid systems

Imre Nagy[1,2], **Áron Süli**[2], **Bálint Érdi**[2]

[1]MTA Research Group for Physical Geodesy and Geodynamics, Műegyetem rakpart 3, 1111 Budapest, Hungary

[2]Department of Astronomy, Eötvös University, Pázmány Péter sétány 1/A, 1117 Budapest, Hungary

E-mail: `i.nagy@astro.elte.hu`

Abstract. In this work we studied the stable regions around four binary asteroids in the main asteroid belt. The studied systems were (107) Camilla, (22) Kallipe, (45) Eugenia and (762) Pulcova. The stability was characterized with three motion indicators: relative Lyapunov indicator, maximum eccentricity, and maximum difference of eccentricities. The survay covered the P type orbits, where satellite moves around both primaries. On the basis of our work it can be decided, in which system the discovery of a third component can be expected.

1. Introduction

The first binary asteroid Ida with its moonlet Dactyl was discovered by the Galileo spacecraft in 1993 (Mason 1994). This binary asteroid orbits in the main asteroid belt which is located roughly between the orbits of Mars and Jupiter. Up to date the number of the main-belt binaries and triplets has grown to 68 and further 36 binary near-Earth objects, and 4 binary Jupiter Trojans have been observed. A list of the actually known systems can be found in `http://www.johnstonsarchive.net/astro/asteroidmoons.html`.

There are three main mechanisms which can form a binary asteroid system. The first is a violent collision. A proto-asteroid can break into two or more parts if it collides with a similar size body. For example the binary asteroid (90) Antiope could be the aftermath of a collision of a 100 km sized proto-Antiope with another Themis family member. This violent shock could have produced the break-up of proto-Antiope into two equisized bodies (Weidenschilling et al. 2001, Marchis et al. 2009). Another mechanism which can form a multiple asteroid system is the gravitional scattering. This mechanism can explain the origin of the Kuiper Belt binaries (Weidenschilling 2002, Goldreich et al. 2002, Funato et al. 2004, Astakhov et al. 2005). At last the so-called thermal YORP (Yarkovsky-O'Keefe-Radzievskii-Paddack) effect can explain the orbital properties of main belt and near-Earth double asteroids (Walsh et al. 2008).

According to the different formation theories, it is very probable that some of the main belt binaries (MBB) are gravitationally bound multiplets, like the Pluto-Charon system in the Kuiper Belt. It might be possible that during the formation process other smaller companions were also formed around them. Actually, two small satellites were discovered around Pluto in 2005 (Weaver et al. 2006). The purpose of this paper is to investigate this possibility. We have explored the dynamical structure of the orbital element space of some selected MBBs (listed in

Table 1. Parameters of selected main belt binaries: a_b is the semi-major axis of the heliocentric orbit of the barycenter; $\mu = m_s/(m_p + m_s)$ is the mass parameter; a_s and e_s are the semi-major axis and the eccentricity of the relative orbit of the secondary with respect to the primary; T_s is the orbital period of the secondary around the primary; R_I is the radius of the sphere of influence of the binary.

Primary	a_b [AU]	Secondary	μ	a_s [km]	e_s	T_s [d]	R_I [km]
(107) Camilla	3.48	S/2001 (107) 1	0.00035	1250	0.002	3.722	16300
(45) Eugenia	2.72	Petit-Prince	0.00005	1180	< 0.002	4.765	9900
(22) Kalliope	2.91	Linus	0.002	1095	< 0.002	3.596	11600
(762) Pulcova	3.16	S/2000 (762) 1	0.002	703	0.03	4.438	8340

Table 1) in order to find, whether there are stable regions in these systems where additional small satellites could exist. Similar investigations were made by Nagy, Süli & Érdi (2006, 2007) for Pluto-Charon system and 7 Kuiper Belt objects.

2. Model, initial conditions and methods

To study the dynamical structure of the orbital element space of the MBBs, we applied the model of the planar elliptic restricted three-body problem. We assumed that the members of a binary, a larger primary and a smaller secondary body, revolve in elliptic orbits around their common center of mass, and a third body of negligible mass, a hypotetical small satellite, moves under the gravitational forces of the two primaries. We investigated the motion of this satellite within the sphere of influence of the binary system. The radius of this sphere is

$$R_I = a_b \left[(m_p + m_s)/m_{Sun}\right]^{2/5},$$

where a_b is the semi-major axis of the orbit of the barycenter of the binary system around the Sun, m_p, m_s, and m_{Sun} are the masses of the primary, the secondary, and the Sun, respectively. Within the sphere of influence, solar perturbations or tides can be neglected, and the members of the binary essentially move around their barycenter in elliptic orbits. Note, that R_I is about 3 times smaller than the Hill radius within which solar perturbations still can be neglected.

The orbital plane of the binary sytem served as reference plane, in which the direction of the pericenter of the secondary's relative orbit with respect to the primary was used as reference direction, from which the angular orbital elements of the satellite were measured. The semi-major axis of the secondary's relative orbit was taken as distance unit, which is denoted with A.

We changed the initial orbital elements of the hypotetical satellite. We determined the stability character of the resulting orbit of the satellite by integrating the equations of motion using a Bulirsch-Stoer integrator and computing three indicators that characterize the dynamical behaviour of the orbit. One quantity was the relative Lyapunov indicator (RLI) (Sándor, Érdi & Efthymiopoulos 2000; Sándor et al. 2004). The RLI measures the difference between the convergence of the finite time Lyapunov indicators to the maximal Lyapunov characteristic exponent of two initially very close orbits. This method is extremely fast in determining the ordered or chaotic nature of individual orbits, as well as to distinguish between ordered and chaotic regions of the phase space. This powerful method has several applications in different problems of planetary dynamics (Érdi et al. 2004; Sándor et al. 2004; Nagy et al. 2006; Sándor et al. 2007).

The second indicator was the maximum eccentricity (ME), reached by an orbit during a given integration time. Orbits with higher eccentricity may become unstable more likely, due to close encounters, thus the ME is a good indicator of stability (Dvorak et al. 2003; Nagy et al. 2006).

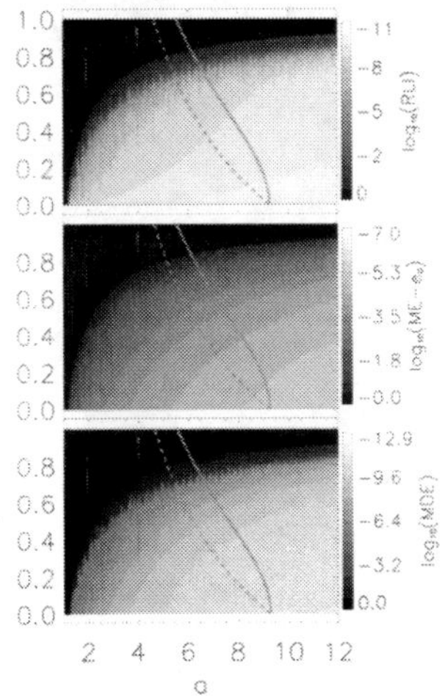

Figure 1. (107) Camilla.

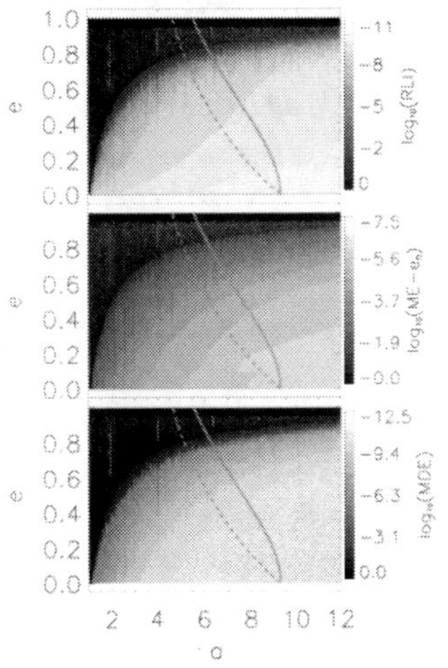

Figure 2. (45) Eugenia.

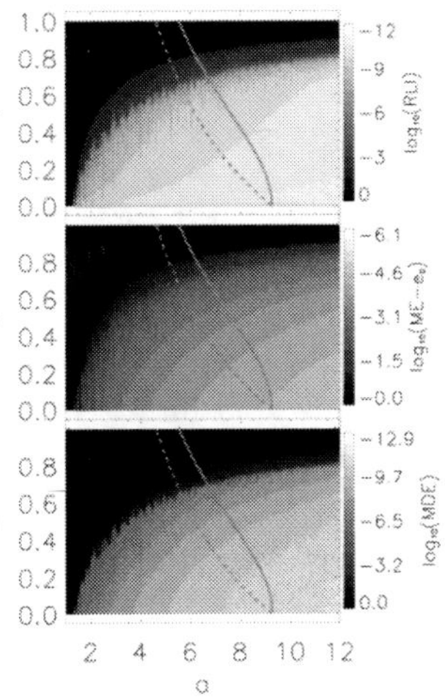

Figure 3. (22) Kalliope.

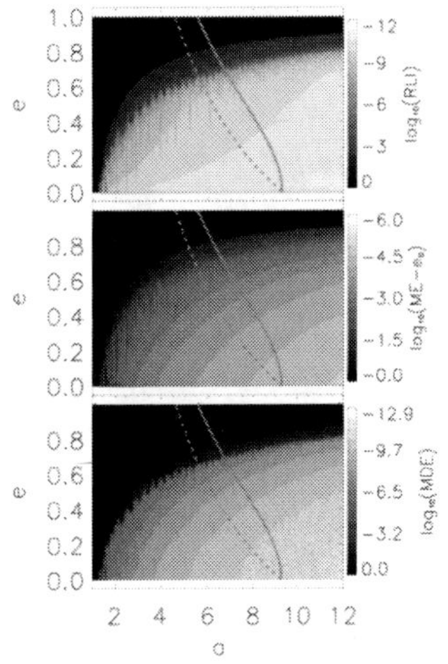

Figure 4. (762) Pulcova.

The third indicator was the maximum difference in the values of the eccentricity (MDE) of two initially very close orbits. This method, introduced by Nagy et al. (2006), is based on the notice that in a chaotic region the eccentricities of two initially close orbits develop quite differently and their momentary differences can be very large even if the average value of the eccentricity of each orbit remains small. The three methods are not equivalent, however, they complete each other.

The integrations were made for 10^3 orbital periods of the secondary (T_s), in order to determine the time span that is long enough to have a trusted value of the indicators. The test runs, performed for all the investigated MBBs, showed that the orbital element space can be surveyed in a reliable way by using a time span of $10^3\,T_s$ for the 4 systems, listed in Table 1.

We plotted the logarithm of the computed indicatoron the plane (a, e). The indicators give information about the dynamical character of the orbits, thus in this way maps of the dynamical properties are obtained.

3. Results

We investigated the stability structure of the parameter space of 4 MBBs listed in Table 1. The results are shown in Figs 1-4. In each panel, the top, middle, and bottom figures show the indicators RLI, ME-e_0, where e_0 is the initial eccentricity and MDE, respectively. Initial points in the light regions correspond to stable orbits, while in the dark domains they result in chaotic motion. These are not necessarily unstable, however, chaotic regions are the birthplaces of unstable motion. In Figs 1-4 a dashed and a solid line are ploted. The orbits right from the dashed line come out from the sphere of influence. If a particle moves in orbit right from the solid line, then it resides outside the sphere of influence in half of its period. In these cases our model assumptions become less and less valid because of the peturbations of the Sun. In these regions the model of the four-body problem should be applied.

(107) Camilla. The parameter space is unstable until $a = 1.5$ A. The boundary line between the stable and unstable regions reaches $e = 0.8$ at $a = 5.5$ A (Fig. 1). The largest stable semi-major axis belong to circular orbit. This is equal to the value of the radius of the sphere of influence, which is 9.3 A.

(45) Eugenia. The stable region begins at $a = 1.3$ A and extends up to $e = 0.73$ at $a = 4.8$ A (Fig. 2). The stable circular orbits extend up to $a = 8.3$ A.

(22) Kalliope. There is instability up to $a = 1.3$ A, and the stable region extends only to $e = 0.68$ at $a = 6.5$ A (Fig. 3). The stable circular orbits extend up to $a = 11$ A.

(762) Pulcova. The stable region begins at $a = 1.3$ A and at $a = 5.6$ A it reaches to $e = 0.65$ (Fig. 4). Circular orbit is stable up to $a = 9.3$ A.

4. Conclusions

The orbits with high eccentricity and large semi-major axes become unstable in every investigated systems within a short time interval. Outside the sphere of influence but inside the Hill sphere orbits can exist with high eccentricity. However, in this region the effect of the Sun becomes significant, consequently a model of the four-body problem should be used stability investigation. Gravitational capturing take place at high eccentric primary orbit, accordingly this mechanism cannot establish a long time stable bounded triplet. However the stable region extends to remarkably large values of the semi-major axes, therefore it is possible that a third body exists in these systems. These bodies might have been born by thermal YORP effect or a major impact.

References

Astakhov S A, Lee E A, Farrelly D 2005 *MNRAS* **360**, 401
Dvorak R, Pilat-Lohinger E, Funk B, Freistetter F 2003 *A&A* **398** L1
Érdi, B, Dvorak R, Sándor Zs, Pilat-Lohinger E, Funk B 2004 *MNRAS* **351** 1043
Funato Y, Makino J, Hut P, Kokubo E, Kinoshita D 2004 *Nature* **427** 518
Goldreich P, Lithwick Y, Sari R 2002 *Nature* **420** 643
Marchis F, Enriquez J E, Emery J P, Berthier J, Descamps P 2009 *American Astronomical Society, DPS meeting no. 41* no.56.10
Mason J W 1994 *Journal of the British Astronomical Association* **104** 108
Nagy I, Süli Á, Érdi B 2006 *MNRAS* **370** L19
Nagy I, Süli Á, Érdi B 2007 *Astronomische Nachrichten* **328** 797
Sándor Zs, Érdi B, Efthymiopoulos C 2000 *CeMDA* **78** 113
Sándor Zs, Érdi B, Széll A, Funk B 2004 *CeMDA* **90** 127
Sándor Zs, Süli Á, Érdi B, Pilat-Lohinger E, Dvorak R 2007 *MNRAS* **375** 1495
Walsh K J, Richardson D C, Michel P 2008 *Nature* **454** 188
Weaver H A, Stern S A, Mutchler M J, et al. 2006 *Nature* **439** 943
Weidenschilling S J, Marzari F, Davis D R, Neese C 2001 *32nd Annual Lunar and Planetary Science Conference, March 12-16, 2001, Houston, Texas, abstract* no.1890
Weidenschilling S J, 2002 *Icarus* **160** 212

Features of librational motions around L_4

Renáta Rajnai[1], **Imre Nagy**[2], **Bálint Érdi**[1]

[1]Department of Astronomy, Eötvös University, Pázmány Péter sétány 1/A, 1117 Budapest, Hungary

[2]Physical Geodesy and Geodesical Research Group of the HAS, Technical University, Műegyetem rkp. 3, 1111 Budapest, Hungary

E-mail: R.Rajnai@astro.elte.hu

Abstract. Trojan bodies are present in the Solar system in great number, as Trojan asteroids and also as Trojan moons. Thus it is possible that their presence is similar in extrasolar planetary systems too. We investigated features of librational motions around L_4 with numerical methods on the mass parameter - eccentricity plane in the elliptic restricted three-body problem. We determined the lifetime of Trojan bodies, until they remained in the librational domain around L_4, also illustrated the distribution of the three possible endgames of trojan motion. Finally we determined the frequences and resonances of the librational motion in the stable region.

1. Introduction

It is not rare in our Solar system, that in a three-body system the third body moves nearly in the same orbit as one of the others, by 60° behind or in front of it. They are in 1:1 mean motion resonance. Trojan asteroids, which are moving co-orbital with Jupiter, are the most known group of configuration of this kind. The first Trojan asteroid, Achilles was discovered in 1906 by Max Wolf. To date astronomers discovered more than 3000 companions of Achilles. We know asteroids at the L_4 and L_5 Lagrangian points of Jupiter nearly 100 years back, while Trojan followers of other planets were detected only in the recent past. By now we know 4 Trojan asteroids of Mars[1] and 6 of Neptune[2]. Trojan bodies can also be moons, namely Saturn has a few such moons[3]. After all, it would not be astonishing, if observations would report about detections of Trojan bodies in extrasolar planetary systems. Configurations of this kind can be studied by using the circular or elliptic restricted three-body problem.

2. The elliptic restricted three-body problem

The recent study investigates a special case of the three-body problem, namely the elliptic restricted three-body problem, when the mass of one body (an asteroid) is negligible, compared to the others, which are moving in elliptic orbits around their common centre of mass. This case can be applied for Trojan asteroids or trojan moons as well.

The mathematical difficulty of this problem is that it can be analysed only by numerical methods. First we investigated the stability time of fictitious trojan bodies by integrating the

[1] Mars: 5261 Eureka, (101429) 1998 VF31, (121514) 1999 UJ7, 2007 NS2.

[2] Neptune: 2001 QR322, 2004 UP10, 2005 TN53, 2005 TO74, 2006 RJ103, 2007 VL305.

[3] Trojan moons: Polydeuces and Helene in the Saturn-Dione system, Telesto and Calypso in the Saturn-Tethys system.

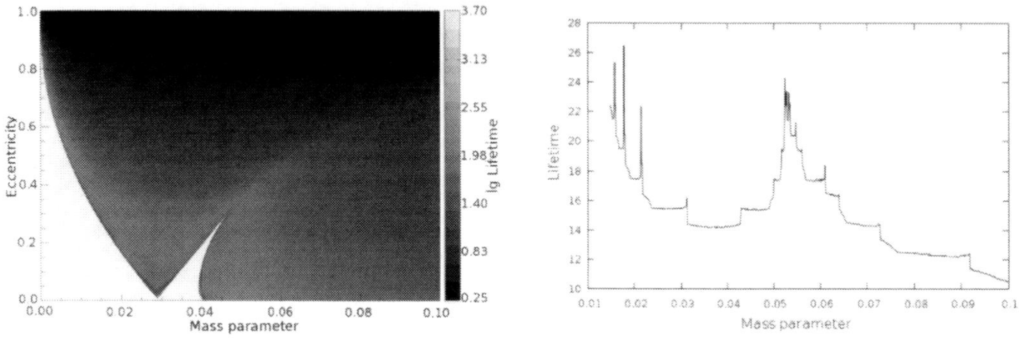

Figure 1. Left panel: The logarithm of the lifetime of orbits with given mass parameter and eccentricity (Érdi et al., 2009). Right panel: Variation of lifetimes as a function of the mass parameter in the unstable region at $e = 0.394$.

equations of motion of the elliptic restricted three-body problem, putting the trojan body in the L_4 point and computing how long it remained there depending on the parameters of the system. P_1 and P_2 marks the two heavier bodies, and P_3 the asteroid, whose mass is negligible. Let P_1 be the centre of the coordinate system, and suppose that P_3 moves in the same plane as the others, so movements in the z direction is neglected. The masses of the components are $1 - \mu$, μ, and 0, respectively, in units of the total mass of the primaries. The initial positions and velocities of P_3 and P_2 were taken such that both begin their motion at the apocentre of their respective elliptic orbit, P_3 being by 60° before P_2.

In this case the equations of motion contain two free parameters: μ (the mass parameter of the system), and the eccentricity e of the elliptic orbit of P_2. Thus we studied the properties of the system on a grid covering the $\mu - e$ plane. We changed the mass parameter μ from 0.0001 to 0.1 with stepsize 0.0001, and the eccentricity from 0 to 1 with stepsize 0.002. This means about 500 000 (μ, e) points.

We also investigated the frequencies of motion around L_4 depending on μ and e. Analysis of the frequencies is based on the definitions of Érdi et al. (2007). To determine the frequencies of the librational motions we used a numerical library for C programs, Gnu Scientific Library. This provides a wide range of mathematical routines, so we used a Fast Fourier Transform (FFT). The description of the algorithm is given in Temperton (1983).

3. Lifetime of the system

We examined the lifetime of the systems with numerical integration through 5000 periods of P_2. We considered the motion of P_3 stable, if it remained in the surroundings of L_4 during the entire time of integration. Also considered it as unstable, if it moved away from P_2 farther than 180° or they became impacted[4]. On the left panel of Figure 1 one can see the stability time of the system at given mass parameter and eccentricity. The plotted value is the logarithm of the lifetime. In the stable region P_3 stays during the entire integration time near L_4 (the region with the bright grey colour, where the logarithm of the lifetime is 3.70), while in the unstable region it remains there only for a few dozen periods (the logarithm of the lifetime is less than 1.40).

The border of the stable region is quite clear, while lifetimes in the unstable region shows a decreasing tendency with some local prominences. This is shown in the right panel of Figure 1

[4] Collision was defined so, that the separation of P_2 and P_3 is smaller, than the Hill radius of P_2.

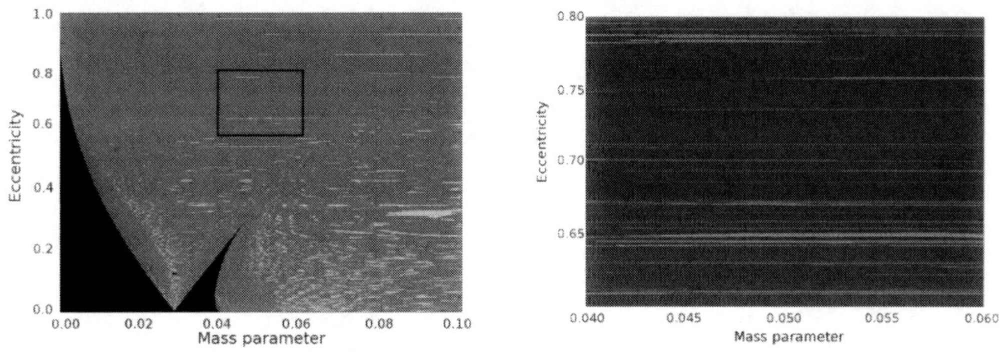

Figure 2. Left panel: Possible motions of P_3. Black: stable motion. Grey: P_3 moved away from P_2 farther than 180°. White: collision occured between P_2 and P_3. Black square: The position of the right panel. Right panel: A part of the left panel, integrated with higher resolution: $\mu \in [0.04; 0.06]$ with stepsize 0.00001, and $e \in [0.6; 0.8]$ with stepsize 0.0001.

for $e = 0.394$. At $\mu = 0.05$ the lifetime increases with a few periods, that can be due to some resonances, but this needs to be investigated more in the future. At higher eccentricities the system dissolves sooner, lifetimes take here just a few periods.

4. The dissolution of the system

The motion of P_3 has the three above-mentioned endgames: a) librational motion through the whole time; b) it moves farther away from P_2 than 180°; c) collision occurs.

On the left panel of Figure 2 is the distribution of the three possible ending of the system. The integration was done here for 5000 periods of P_2 also. The black region shows the systems, in which the motion of P_3 is stable during the whole integration time, grey colour is for ejection from the librational domain, and the white fine fibrous structure is where P_2 and P_3 collide. It can be seen, that at smaller eccentricities the system dissolves usually because of collision, and at higher eccentricities ejection from the librational domain dominates the $\mu - e$ plane. Near to the border of the stable region, the third body with the very small mass usually moves farther away than 180° from P_2, opposite to the $\mu > 0.05$ domain, where colliding systems form fine fibrous structures, usually by constant eccentricities. This can look like integrational error, but at higher resolution they become thicker, and beside them several new thin structures are also appearing (Figure 2, right panel).

5. Frequencies and resonances

To determine the frequencies of the librational motion of P_3 we used Fourier transform, and integrated for 1250 periods of P_2 to get practicable Fourier spectra. But this method could not map the unstable region, because at ejection from the librational domain the spectrum has exponential shape, no peak could be identified, and in the case of collision the stability time was too small to get a practicable Fourier spectrum.

Figure 3 shows the variation of the four frequencies of the librational motion (n_s, $1 - n_l$, n_l, $1 - n_s$, corresponding to the elliptic restricted three-body problem, n_s, n_l being the normalized frequences of short and long period libration) as the function of the mass parameter. The lack of data around $\mu = 0.03$ is due to the instabil region, where we could not get the frequences with FFT.

Knowing the frequencies of the librational motion, we can determine resonances. Resonances can appear when the ratios of frequencies are rational numbers, so they can be given as the

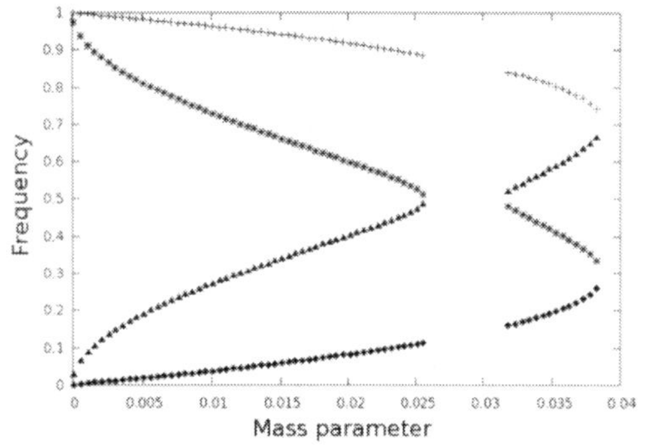

Figure 3. Variation of the frequencies at $e = 0.01$ as the function of the mass parameter. Key: cross: n_s, star: $1 - n_l$, triangle: n_l, diamond: $1 - n_s$ frequencies.

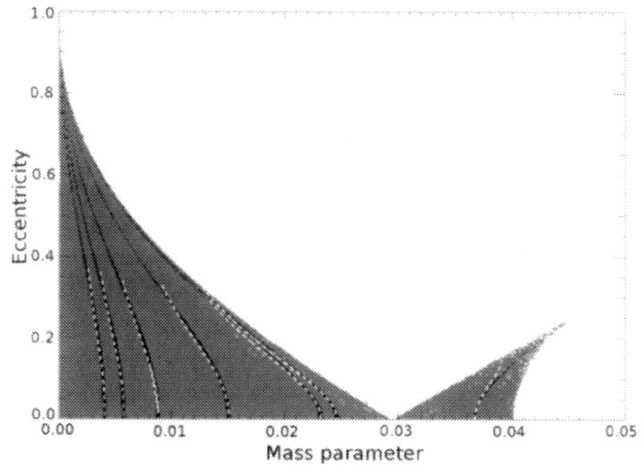

Figure 4. The $(1 - n_l) : n_l$ resonances. Black thick lines: resonances derived numerically, from left to right: 5:1, 4:1, 3:1, 2:1, 4:3, 5:4, 2:3. White dashed lines are the same resonances derived with analytical methods.

ratios of integer numbers. We used this fact to find resonances on the $\mu - e$ plane. For the four frequencies of the elliptic restricted three-body problem 6 types of resonances exist (Érdi et al. 2007, 2009). Figure 4 shows one of them: $(1 - n_l) : n_l$. The grey region is where the system is stable, black thick lines denotes the resonances obtained by numerical integration, and the dashed lines presents resonances determined by analytical methods. The latter was computed by Érdi et al. (2007), using Rabe's equation (Rabe, 1973). It can be seen that resonances, determined by different methods, fit very well at smaller eccentricities. Of course the method of Rabe can be used only for small and moderate values of e, that is why the curves differ at higher eccentricities. The turning points can denote the validity limit of Rabe's equation.

6. Summary

We investigated the lifetime, until the asteroid with negligible mass remained in the three-body system, for the $\mu - e$ plane (Figure 1), and found, that this plane can be divided into a stable and unstable region. For the previous case P_3 remained in the surroundings of L_4, and the latter led to dissolution. Lifetime of the system decreases quickly in the unstable region, as the eccentricity increases, or far from the boundary of the stable domain.

Distribution of the three possible outcomes were determined, this is shown in Figure 2. We found that at smaller eccentricities dissolution is due to collision, and as e increases, ejection from librational domain becomes dominant.

Furthermore we determined frequencies and resonances for the stable librational motions, and found that resonance, obtained with numerical integration fit in with that, obtained with analytical methods (Figure 4). Future work could be to determine the characteristic roots of the motions in the whole $\mu - e$ plane, thus get the possible frequencies and resonances in the unstable region.

7. References

Érdi B, Nagy I, Sándor Zs, Süli Á, Fröhlich G 2007 *Monthly Notices of the Royal Astronomical Society.* **381** 33
Rabe E 1973 *Recent Advances in Dynamical Astronomy* (D. Reidel Publ. Co., Dordrecht) p 156
Temperton C 1983 *Journal of Computational Physics* **52** 340
Érdi B, Forgács-Dajka E, Nagy I, Rajnai R 2009 *Celestial Mechanics and Dynamical Astronomy* **104** 145-158

5th Workshop of Young Researchers in Astronomy & Astrophysics IOP Publishing
Journal of Physics: Conference Series **218** (2010) 012019 doi:10.1088/1742-6596/218/1/012019

Stable and habitable systems with two giant planets

Vera Dobos[1], **Imre Nagy**[2,1] **and Judit Orgoványi**[1]

[1] Eötvös University, Pázmány Péter sétány 1/A, 1117 Budapest, Hungary
[2] Physical Geodesy and Geodesical Research Group of the HAS, Technical University, Műegyetem rakpart 3, 1111 Budapest, Hungary

E-mail: vera.zs.dobos@gmail.com

Abstract. We have studied planetary systems which are similar to the Solar System and built up from three inner rocky planets (Venus, Earth, Mars) and two outer gas giants. The stability of the orbits of the inner planets is discussed in the cases of different masses of the gas planets. To demonstrate the results stability maps were made and it was found that Jupiter could be four times and Saturn could be three times more massive while the orbits of the inner planets stay stable. Similar calculations were made by changing the mass of the Sun. In this case the position of the rocky planets and the extension of the liquid water habitable and the UV habitable zones were studied for different masses of the Sun. It was found that the orbits of the planets were stable for values greater than 0.33 $M_\odot$ where $M_\odot$ is the mass of the Sun and at lower masses of the Sun (at about 0.8 $M_\odot$) only Venus, but for higher mass values (at about 1.2 $M_\odot$) Earth and also Mars are located in both habitable zones.

1. Introduction

New planetary systems containing one or more gas giants are discovered day by day. Present instruments and methods are, however, not capable of detecting rocky planets, hence these discoveries are considered to be rare events. As instruments are getting more and more developed the number of known terrestrial-mass planets is expected to grow significantly (Csizmadia, 2009).

Search for extraterrestrial life is also one of the main goals of ESA and NASA. New interferometers of these space agencies are to be launched in the next decade with the aim of discovering and characterizing new Earth-like planets. Therefore it is important to investigate the conditions within which the orbits of these planets are stable in the given system.

What would be the outcome in terms of the stability of the Solar System, if the masses of Jupiter and Saturn had been formed in a different way from the cloud envelope of the protostellar Sun? Is the Solar System unique? In this work we discuss different cases similar to the Solar System to investigate the conditions of habitability on rocky planets.

Our research program consists of two main parts. The first part focuses on calculations related to celestial mechanics that is the stability of the rocky planets is investigated as a function of the masses of the gas giants and as a function of the mass of the Sun. This latter case is closely related to the second part of our work which is the calculation of the extension of the two kinds of habitable zones: the conventional liquid water habitable zone and the ultraviolet habitable zone.

2. Methods

2.1. Stability calculations

The studied planetary systems are built up from three inner rocky planets: Venus, Earth, Mars and two outer gas giants: P_1 and P_2, where P_1 is closer while P_2 is located farther from the Sun. M_1 and M_2 are the masses of P_1 and P_2, respectively. By changing the masses of the gas giants from 0 to 10 M_J where M_J is the mass of Jupiter the integration was made for 5000 P_S (P_S: period of Saturn) which is about 150000 years. Initial orbital elements were obtained for J.D. = 2415020.0 epoch where P_1 and P_2 were chosen for the parameters of Jupiter and Saturn, respectively. Stability of the inner planets is described by the maximum of eccentricities and close encounters. For the latter we investigated several parameters but finally chose the difference between the distance of the outer planet's pericentre and the distance of the inner planet's apocentre.

In the other case the planets' masses were constants and the mass of the Sun was altered between 0.1 $M_\odot$ and 10 $M_\odot$ where $M_\odot$ is the mass of the Sun. The integration time was the same as in the previous case that is 5000 P_S. The stability was characterized by the pericentre and apocentre of the planets and both habitable zones were calculated for a few masses. For the latter the radius of the star was calculated from its mass using the formulae constructed by Zaninetti (2008).

2.2. Habitable zones

The liquid water habitable zone (LW HZ) is a region around a star in which an Earth-like planet could support liquid water on its surface (Kasting et al., 1993). In the equation that describes the boundaries of the LW HZ the Solar system is used as a reference for the calculations of other planetary systems.

An idea from Buccino et al. (2006) was borrowed that there must be both an inner and an outer boundary for sufficient and necessary UV radiation level respectively (UV HZ). It is well known that this radiation induces DNA[1] damage (especially UVB and UVC), inhibits photosynthesis (UVA) and causes lesion in a wide variety of proteins and lipids (Cockell, 1998). The action spectrum $B(\lambda)$ is commonly used as a measure of damage, quantifying the injurious effects of the UV radiation on the biological processes in function of the wavelength (Coohill, 1991; Horneck, 1995; Cockell, 1998). The action spectrum is determined by exponential functions given by Modos et al. (1999) who measured the spectrum by different dosimeters. The results of uracil dosimeter were chosen for the calculations as uracil is a compound that can be found within the RNA[2].

The biological effective spectrum for each value of wavelength can be obtained by multiplying the action spectrum with the incident radiation. To get a tractable expression for the calculation of the inner boundary, the Earth was used again as a reference. This way the following inequality was obtained:

$$\frac{d_{inner}^2}{d_\oplus^2} \geq \frac{\int_{200nm}^{400nm} B(\lambda) \cdot R_*^2 \cdot E_*(\lambda)\, d\lambda}{x \cdot \int_{200nm}^{400nm} B(\lambda) \cdot R_\odot^2 \cdot E_\odot(\lambda)\, d\lambda}, \tag{1}$$

where d_{inner} and $d_\oplus$ are the distances of the inner boundary and of the Earth, R is the radius and $E(\lambda)$ is the emitted energy of the star and of the Sun at λ wavelength and x is the multiplicator of the terrestrial UV radiation that still can be tolerated by DNA, proteins and photosynthetic

[1] The deoxyribonucleic acid (DNA) contains the genetic instructions used in the development and functioning of all known living organisms.

[2] The ribonucleic acid (RNA) conveys genetic information and catalyzes important biochemical reactions.

process. In other words this means that these important compounds can bear even x times of the UV radiation on Earth.

On the other hand the UV radiation is an important energy source needed for chemical synthesis of complex molecules. An expression similar to (1) can be formulated to describe the outer boundary of the UV habitable zone:

$$\frac{d_{outer}^2}{d_{\oplus}^2} \le \frac{\int_{200nm}^{400nm} R_*^2 \cdot E_*(\lambda)\, d\lambda}{y \cdot \int_{200nm}^{400nm} R_{\odot}^2 \cdot E_{\odot}(\lambda)\, d\lambda}, \tag{2}$$

where d_{outer} is the distance of the outer boundary and y is the multiplicator of the terrestrial UV radiation which is absolutely necessary for chemical reactions. Obviously y should be set to be less than 1.

The borders of the UV HZ can also be calculated for the ancient Sun since its radiation has increased by about 30% in the last $4.7 \cdot 10^9$ years (Gough, 1981). In this work the present radiation level was used for the calculations.

3. Results and discussion

3.1. Stability calculations

Stability is investigated by calculating the extrema of the orbital elements (semimajor axis, eccentricity, pericentre and apocentre distance). Using these data stability maps in function of the giant planets' masses (M_1 and M_2) were plotted (see Fig. 1 (*a*), (*b*) and (*c*)). Axis x corresponds to the mass M_1 while axis y to the mass M_2 (both in units of M_J). Yellow coloured regions contain those pairs of masses for which the orbits of the rocky planets are the most stable while black colour indicates the most unstable cases. One can see the '+' mark at coordinates (1, 1/3) which corresponds to one Jupiter and one Saturn mass, so this is the case of our Solar System. It can be observed that the figures are almost symmetrical to the 45° diagonal line. This is not surprising regarding that the masses of the two giant planets are expressed in the same unit. A slight asymmetry can also be noticed, that is caused by the different distances of the giant planets. For all inner planets the maximum eccentricity is extremely high (black colour) for low M_1 mass when M_2 is about $9-10\ M_J$. The reason is that in these cases M_1 gets closer to the Sun, at about 3 AU, thus the inner planets migrate outward. Other conspicuous characteristics are the horizontal lines at $1-2.5\ M_J$ of M_2 at high masses of M_1. The explanation is that the maximum semimajor axis of Venus gets higher and thus the planet gets closer to Earth. Even in some cases the orbit of Venus partially overlaps the orbit of Mars. If we take a look at the stability map of Mars a big black area is clearly seen at the top right corner. This unstable region is formed by the influence of the two massive giant planets. Here the maximum eccentricity of Mars almost reaches the value of 1.

The planet is considered unstable if the difference between its maximum and original eccentricity is greater than 0.2. Fig. 1 (*d*) is a composition of the previous stability maps. It shows the number of stable inner planets in function of the masses of the giant planets. Yellow colour means that all three planets are considered stable according to the above criterion while black indicates unstable regions (when all three planets are unstable). The yellow region has a very large extension which means that the orbits of the inner planets are stable for a wide variety of the pairs of the gas giants.

Stability is not only determined by maximum eccentricities, but close encounters were also examined. To characterize the vicinity of neighbouring rocky planets the difference between the locations of these planets was calculated at every integration step. From these values the smallest was chosen as a measure of the encounter (D). To get comparable results for the two

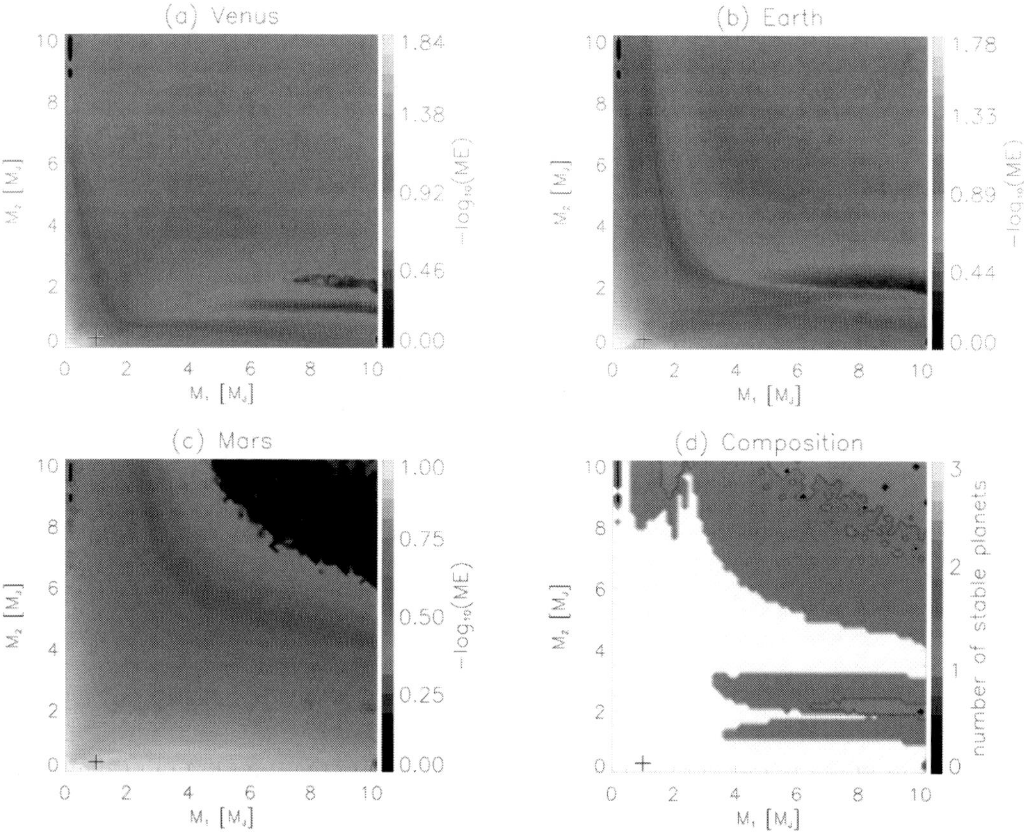

Figure 1. (a), (b) and (c): Stability maps for the three inner planets in function of the masses of M_1 (axis x) and M_2 (axis y). ME: maximum value of the eccentricity. Yellow and black colours indicate the most stable and the most unstable regions, respectively. The '+' mark indicates our Solar System. (d): Composition of stability maps seen in (a) – (c). Colours indicate the number of stable planets: yellow - three, red - two, green - one, black - zero.

planet pairs (*Venus* – *Earth* and *Earth* – *Mars*) these values were normalized by the mean of the original semimajor axes:

$$D = \frac{min\{encounters\}}{\frac{a_1+a_2}{2}}, \tag{3}$$

where a_1 and a_2 are the semimajor axes of the closer and the farther planet from the Sun. The value of D for different mass-pairs of giant planets can be seen in Fig. 2 (a) and (b). Yellow and black colours indicate the least close and the closest encounters, respectively. The *yellow* – *orange* boundaries can be considered as the margin of instability. In the case of Venus and Earth encounters are only seen for mass values $M_1 \geq 6\ M_J$ and $M_2 \approx 2\ M_J$. But at Earth and Mars at the top right corner (at great masses of the giant planets) a huge black region can be seen caused by the instability of Mars mentioned above. In this area in several cases the semimajor axis of Mars is about 1 – 2 AU and its eccentricity almost reaches the value of 1.

From these stability maps a composite figure was made (see Fig. 2 (c)). The planet pair is considered unstable if the ratio of the value of the encounter and the mean of the original semimajor axes is less than 0.2. In Fig. 2 (c) the number of stable planet pairs can be

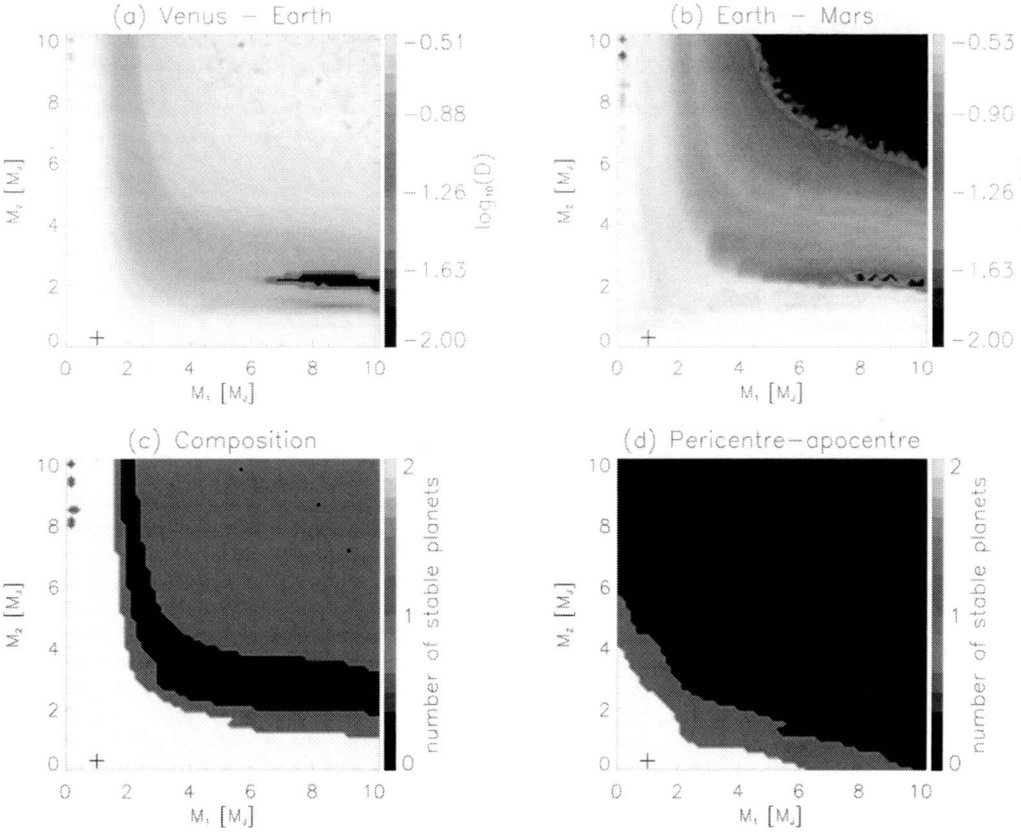

Figure 2. (a) and (b): Stability maps from close encounters. Yellow and black colours indicate the most stable and the most unstable regions, respectively. (c): Composition of stability maps seen in (a) and (b). (d): Stability map made from the difference between the minimum pericentre of the outer planet and the maximum apocentre of the inner planet. Colours indicate the number of stable planet pairs: yellow - two, brown - one, black - zero. The '+' mark indicates our Solar System.

seen. Yellow and black colours indicate the case when both and neither planet pairs are stable, respectively. To characterize the vicinity of the orbits of neighbouring rocky planets the difference between the outer planet's minimum pericentre distance and the inner planet's maximum apocentre distance was calculated as well. The orbits of the planet pairs are considered unstable if the ratio of the difference mentioned above and the mean of original semimajor axes is less than 0.01 (i.e. the planets come very close to each-other). In Fig. 2 (d) the number of planet pairs with stable orbits can be seen. This is the worst case that can be imagined, when the outer and the inner planets are located in their pericentre and apocentre, respectively, and these points and the Sun lie in a line. This may occur but the most probable case may be somewhere between the cases depicted in panel (c) and (d). In panel (d) yellow colour indicates that the orbits are surely stable while in panel (c) it indicates that they may be stable. Even in the worst case the stable (yellow) region is considerably large, hence Jupiter could be four times and Saturn could be three times more massive while the orbits of the inner planets stay stable.

The other problem that was investigated is also related to the stability of the inner planets but in this case the masses of the gas giants were constants while the mass of the Sun was altered between 0.1 $M_\odot$ and 10 $M_\odot$. The minimum pericentre and the maximum apocentre distances of the planets are plotted in Fig. 3 in function of the mass of the Sun. At low star masses the orbits of the inner planets can overlap each-other, hence two or even all three planets can encounter. From 0.33 $M_\odot$ (vertical spotted line on the figure) the orbits of the planets are considered stable regarding to the above-mentioned criterion (i.e. the ratio of the value of the encounter and the mean of the original semimajor axes of Venus and Earth is greater than 0.01).

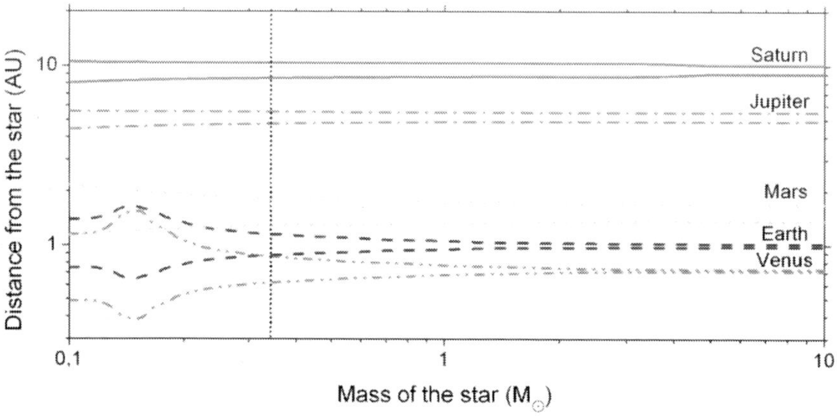

Figure 3. The minimum pericentres and the maximum apocentres of the planets in function of the mass of the star. Different colours indicate different planets. The vertical spotted line at 0.33 $M_\odot$ is a limit for the stability: for less mass values the orbits of the planets are considered unstable and for greater values the orbits are considered stable.

In order to better understand the feature of habitability the two kinds of habitable zones were calculated. The results of these investigations are to be thoroughly discussed in the next chapter.

3.2. Habitable zones

The boundaries of habitable zones were calculated for seven values of mass in the 0.8 – 1.2 $M_\odot$ interval using the definition of the LW HZ given by Kasting et al. (1993) and equations (1) and (2) for the UV HZ. Two x,y sets were used for calculating the boundaries of the UV HZ. The first was $x = 2$ and $y = 1/2$ which was used by Buccino et al. (2006) as well. These are estimated values since different compounds and organisms have different level of tolerance (Cockell, 1998). As the values of these constants are difficult to estimate the extension of the zone was also investigated in the case of $x = 3$ and $y = 1/3$. The extension of both habitable zones for the seven masses of the star can be seen in panel (a) and (b) of Fig. 4. In the first case ($x = 2$ and $y = 1/2$) the UV HZ has a similar extension to the conventional HZ and they overlap each-other, but the UV HZ is located slightly closer to the Sun in the 0.8 – 1.2 $M_\odot$ interval. In the other case the UV HZ covers a wider range than the conventional HZ, especially the outer boundary moved farther. For higher masses than 0.95 $M_\odot$ the LW HZ is fully located inside the UV HZ. It can be seen that the Earth is located in both habitable zones even for lower and higher masses of the Sun.

For detailed results see Fig. 5, where black and light blue colours indicate again the conventional and the UV ($x = 2$ and $y = 1/2$) habitable zones, respectively. At lower masses of the Sun only Venus, but for higher mass values Earth and also Mars are located in both HZ.

Journal of Physics: Conference Series **218** (2010) 012019 doi:10.1088/1742-6596/218/1/012019

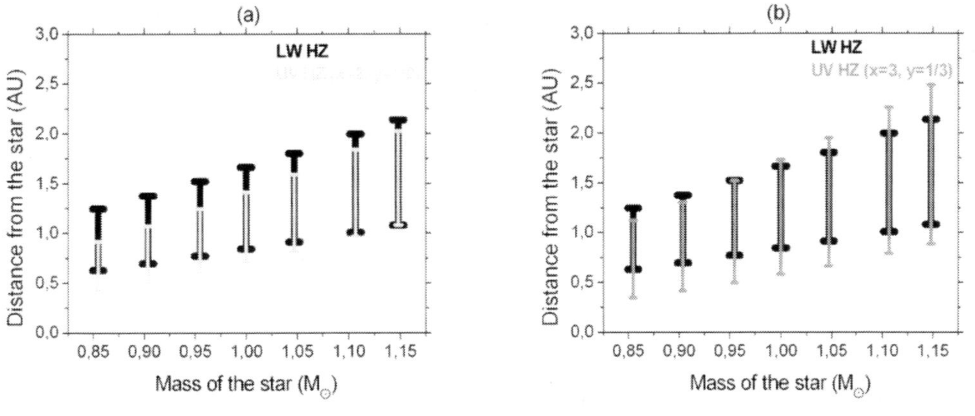

Figure 4. The extension of the habitable zones in function of the mass of the star. (*a*): Black and light blue lines indicate the conventional and the UV habitable zone ($x = 2$, $y = 1/2$), respectively. (*b*): Black and orange lines indicate the conventional and the UV ($x = 3$, $y = 1/3$) habitable zones, respectively.

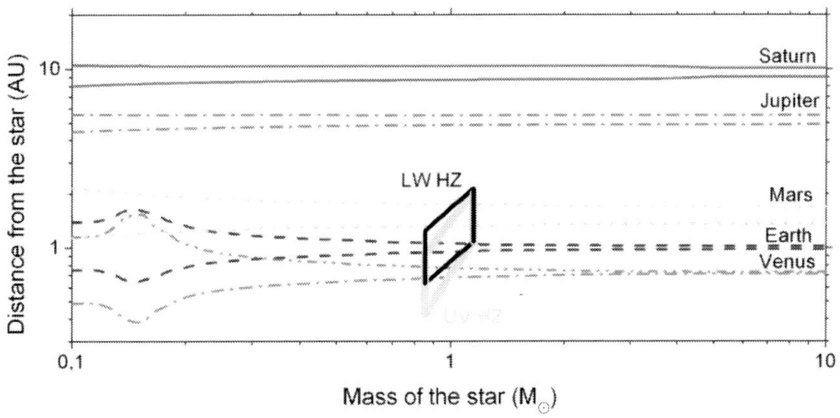

Figure 5. The minimum pericentres and the maximum apocentres of the planets like in Fig. 3. Between 0.8 and 1.2 $M_\odot$ the extension of the habitable zones can be seen: black and light blue lines indicate the conventional and the UV habitable zone ($x = 2$, $y = 1/2$), respectively.

4. Conclusions

In the celestial mechanics discussion the stability of the orbits of the rocky planets was investigated for different masses of the two gas giants. The inner planetary system remained stable for several mass pairs, hence the fine-tune problem does not prevail. Yet, long-term investigation would be worthwhile because Batygin and Laughlin (2008) found that Mars can escape from the Solar System in 800 – 1200 million years.

In the other case when the mass of the Sun was altered, the orbits of the planets were stable for values greater than 0.33 $M_\odot$. Between 0.8 and 1.2 *AU* the extensions of the two habitable zones were calculated and in almost all cases the Earth was located in both habitable zones. This result also suggests that the Solar System is not considered to be unique.

Acknowledgments
We thank Dr. Bálint Érdi for useful comments and discussion.

References
Batygin K and Laughlin G 2008 On the dynamical stability of the Solar System *ApJ* **683** 1207-16

Buccino A, Lemarchand G and Mauas P 2006 Ultraviolet radiation constraints around the circumstellar habitable zones *Icarus* **183** 491-503

Cockell C 1998 Biological effects of high ultraviolet radiation on early Earth - a theoretical evaluation *J. Theor. Biol.* **193** 717-29

Coohill T P 1991 Action spectra again? *Photochem. Photobiol.* **54** 859-70

Csizmadia S 2010 Notes on exoplanets *J. Phys.: Conf. Series* in this volume

Gough D O 1981 Solar interior structure and luminosity variations *Solar Physics* **74** 21-34

Horneck G 1995 Quantification of the biological effectiveness of environmental UV radiation *J. Photochem. Photobiol.* B: Biol. **31** 43-9

Kasting J, Whitmire D and Reynolds R 1993 Habitable zones around main sequence stars *Icarus* **101** 108-28

Modos K, Gaspar S, Kirsch P, Gay M and Ronto G 1999 Construction of spectral sensitivity function using polychromatic UV sources *J. Photochem. Photobiol.* B: Biol. **49** 171-6

Zaninetti L 2008 Semi-analytical formulas for the Hertzsprung-Russel Diagram *Serbian Astronomical Journal* **177** 73-85

How 'New Horizons' will see the Pluto-Charon system

Zsuzsa Zsigmond[1] **and Áron Süli**[1]

[1]Department of Astronomy, Eötvös University, Budapest, Hungary

E-mail: `zs.zsigmond@astro.elte.hu`

Abstract. We present a detailed survey of the dynamical structure of the phase space around the new moons of the Pluto–Charon system. We investigated the system in the framework of the spatial elliptic restricted three-body problem. Stability maps were created using chaos indicators both on the semimajor axis - eccentricity and semimajor axis - inclination plane. The structures related to the 4:1 and 6:1 mean motion resonances are clearly visible on the maps, but the detailed investigation of the resonant arguments shows no evidence of the resonances. We showed the possibility that Nix might be in the 4:1 resonance if its argument of pericenter or longitude of node falls in a certain range. Recently we improved our model to the general four-body problem, which can provide more accurate information about the stability, motion and possible resonances between the moons and Charon, which is relevant, since the 'New Horizons' mission will reach and study the system in 2015.

1. Introduction

Pluto's first moon, Charon was found by Christy & Harrington (1978), which greatly facilitated the study of Pluto. The Pluto–Charon system is remarkable, since in the Solar system Charon is the largest moon relative to its primary, with the highest mass ratio of 0.1166 (Tholen 2008, hereafter referenced to as T08).

The discovery of Pluto's new moons (Weaver 2005, Weaver 2006), Nix and Hydra rendered the system even more interesting. Thank to this, in latter years numerous studies were devoted to the system.

In 2006 two-body orbit solutions for Nix and Hydra were computed by Buie using the Hubble Space Telescope (Buie 2006, hereafter referenced to as B06). According to the results, the orbital periods of Nix and Hydra are close to the ratio of 4:1 and 6:1 with that of Charon, respectively, indicating mean-motion resonances. In this paper two bodies are in mean-motion resonance, when $n/n' = p/(p+q)$, where n and n' are the mean motions, p and q are small integers, where q is the order of the resonance. Some of Buie's major conclusions are: (*i*) Nix and Hydra are in nearly circular orbits, with eccentricities of 0.0023 and 0.0052, respectively, (*ii*) the orbits of the small moons are almost coplanar with Charon's orbit, (*iii*) the orbital periods of Nix and Hydra are nearly commensurate with the period of Charon, but differ significantly from the exact ratios of 4:1 and 6:1, respectively. We note that in B06 the eccentricity of Charon was assumed to be zero. We demonstrated that the eccentricity of Charon substantially influences the phase space of the Pluto–Charon system; for details see Süli & Zsigmond (2009, hereafter referenced as S09).

In 2008 four-body orbit solution was computed by T08 fitting all the 22 parameters simultaneously. They studied whether the direct perturbations might be too strong to permit a sufficiently accurate extrapolation forward to the '2015 - New Horizons' spacecraft encounter with Pluto. Also, with adequate data, this four-body orbit solution should yield the mass for each member of the system. Their major conclusions are the follows: (*i*) Pluto's three moons are not quite coplanar, (*ii*) the new moons' orbit planes precess around the system's invariable plane with periods of ≈ 5 and ≈ 15 years, (*iii*) the orbital eccentricities are nonzero, but small for all three moons, (*iv*) there is no evidence of any mean motion resonances.

Nagy et al. (2006) studied the phase space of the Pluto–Charon system in the framework of the spatial circular restricted problem. The moons were treated as test particles and their semimajor axes, eccentricities and inclinations were varied. Summarizing their results: (*i*) the region inside 42 000 km is unstable, thus no moon could exist there, (*ii*) both moons reside in the stable region and the upper limit for the eccentricities are 0.17 for Hydra and 0.31 for Nix, far greater, than the current values of the moons, (*iii*) in the semimajor axis–inclination plane the 4:1 and 6:1 resonances are clearly visible above $\approx 20°$ and $\approx 35°$, respectively. Our main goal was to extend the previous investigations using the spatial elliptic restricted and general four body problems. Detailed analysis can be found in the work of S09. This paper is organized as follows: Section 2 describes the model and the initial conditions; Section 3 is devoted to the methods and in Section 4 we present the new results.

2. Model and initial conditions

Since the orbital radii of the moons are much smaller than the Hill radius ($\approx 8.0 \times 10^6$ km) of the Pluto-Charon system, the moons are deep in Pluto's gravitational well, so the perturbations by the Sun can be ignored, as did T08.

First we applied the model of the spatial elliptic restricted three-body and later the general four-body problem. We integrated the dimensionless equations of motion. An obvious advantage of using such equations is that the results are independent of the exact value of the semimajor axis of Charon. The unit of length was chosen such that the separation of Pluto and Charon (the primaries) is unity, i.e. the semimajor axis of Charon $a_1 = 1$ A in all computations.

Let the unit of mass be the sum of the primaries, and the unit of the time the orbital period of Charon. The orbital plane of the primaries was used as reference plane, in which the line connecting the primaries at $t = 0$ defines a reference x-axis. The initial orbital elements are given in S09.

We study the problem more generally by considering the effect of non-zero inclinations on the orbital stability. The mass parameter $\mu = 0.104424$ was chosen according to the mass ratio of 0.1166, published in T08. To examine the phase space in the vicinity of Nix and Hydra separately we varied the initial orbital elements of the test particles. Stability maps were created for the $(a - e)$ and $(a - i)$ orbital element space for both moons for different eccentricity e, inclination i, argument of pericenter ω and longitude of node Ω values (for details see S09).

3. Methods

To compute the stability maps, the method of the maximum eccentricity (ME), the Lyapunov characteristic indicator (LCI) and the relative Lyapunov indicator (RLI) were used as tools for stability investigations of the massless bodies representing the small moons.

The ME method uses as an indication of stability a straightforward check based on the eccentricity. This action-like variable shows the probability of orbital crossing and close encounter of two bodies and therefore its value provides information on the stability of orbits. This simple check was found to be a powerful indicator of the stability character of orbits in previous investigations (Dvorak et al. 2003; Süli et al. 2005; Nagy, Süli & Érdi 2006). In this work we define ME as follows:

$$\mathrm{ME} = \max_{t\in[0,10^4]T_C}(e).$$

We plotted ME_-, where:

$$\mathrm{ME}_- = \begin{cases} 1 & \text{if ME} = 1 \\ \mathrm{ME} - e_0 & \text{otherwise} \end{cases},$$

where e_0 is the initial eccentricity of the test particle.

As a complementary tool, we computed also the RLI (Sándor et al. 2006) and the LCI (Froeschlé 1984), which are well-known chaos indicators.

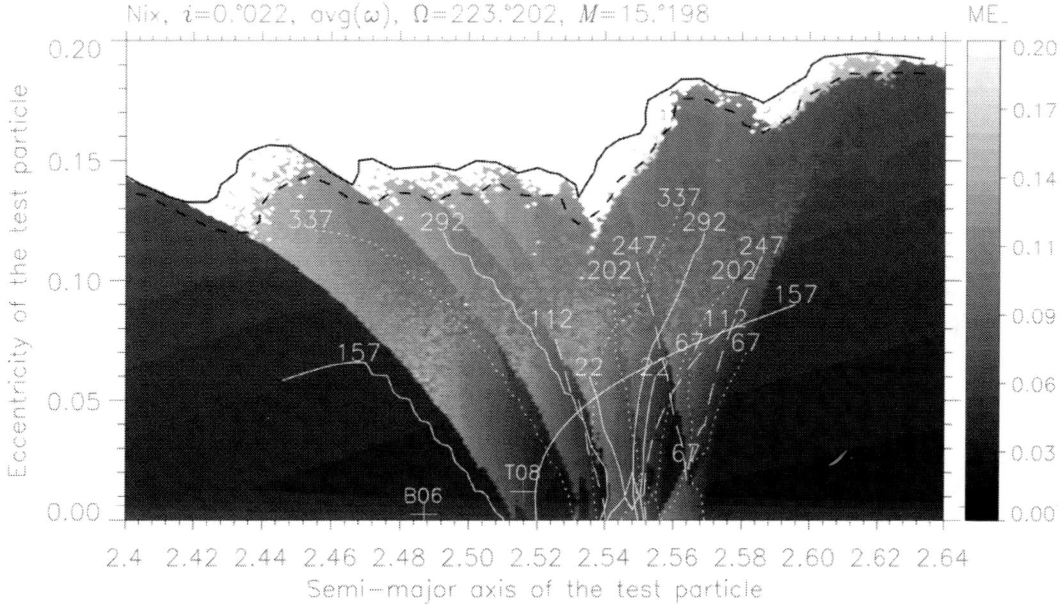

Figure 1. The $(a{-}e)$ stability map for 8 ω values for Nix. Along the contour curves $\mathrm{ME}_- = 0.04$. The location of the plus signs corresponds to the orbital elements published by B06 and T08.

4. Results

Here we show the ME_- stability maps, presenting one phase space structure for Nix. On Figure 1 white solid lines are contour lines: along them ME_- has a constant value, this way they can draw the approximate boundary of the most prominent structures. The white numbers are the corresponding ω values.

4.1. Mass parameter

We performed two different runs to estimate the effect of the different values of the mass parameter used by Nagy et al. (2006) and by T08. The structures for the two mass parameter are almost identical, for $\mu = 0.104424$ they are only shifted along the horizontal axis. This is a consequence of the changes in the coordinates of the barycenter. This shift could be important in the close vicinity of mean motion resonances.

4.2. Charon's eccentricity

In order to visualise the effect of the absence of the Jacobi-integral, we plotted also the circular and the elliptic case for both moons. The differences are: in the elliptic case the unstable zone is much larger and the center of the resonance is shifted to larger semi-major axis, and the 4:1 and 6:1 resonances became stronger for low eccentricities. The V-shape did not change significantly.

4.3. The stability maps

We investigated the orbits systematically by changing the initial orbital elements of the test particle. In Figure 1, the results are summarized for the 8 values of ω. The figure is dominated by a V-shaped gray structures, corresponding to the 4:1 mean motion resonance between Charon and Nix. These resonances can represent either ordered or weakly chaotic behaviour. The present positions of the moons are in stable region both on the $(a - e)$ and $(a - i)$ parameter spaces. On the maps the structures related to the 4:1 and 6:1 resonances are clearly visible, but none of them contains any of the moons. As you can see in Figure 1, in the case of $\omega = 157.9$ for Nix it can show the possibility of active resonance. But for Hydra we can not find any value, where the 6:1 resonance would appear, since it is very weak in the vicinity of the moon. The $(a - i)$ maps give the same results. Mention must be made that the planar and spatial cases are almost identical, when the inclinations are very small (as they are for the moons).

4.4. Results using the general problem

Newly we started to investigate the system using the framework of the general four-body problem. The masses and their error limits were from work of T08. Using the maximal margin of error, our initial results show that the moons are on orbits those very hard to estimate for larger time. First we calculated the well-known Lyapunov-time (T_L). The Lyapunov time reflects the limits of the predictability of the system. By convention, it is measured as the time for nearby trajectories of the system to diverge by e (Euler-number). We found that for the new moons $T_L = 20$ day. Since the 'New Horizons' mission will reach and study the system in 2015 (Stern 2002), it is very relevant to know the precise positions of the system's members. This claims further investigations which are under way.

References

Benettin G, Vecchio G L and Tenenbaum A 1980 *Phys. Rev. A* **22** 1709
Buie M W, Grundy W M, Young E F, Young L A and Stern S A 2006 *The Astr. J.* **132** 290
Christy J W and Harrington R S 1978 *The Astr. J.* **83** 1005
Dvorak R, Pilat-Lohinger E, Funk B and Freistetter F 2003 *Astr. & Astroph.* **398** L1
Froeschlé Cl 1984 *Cel. Mech.* **34** 95
Nagy I, Süli Á and Érdi B 2006 *MNRAS* **370** L19
Sándor Zs, Süli Á, Érdi B, Pilat-Lohinger E and Dvorak R 2006 *MNRAS* **375** 1495
Süli Á, Dvorak R and Freistetter F 2005 *MNRAS* **363** 241
Süli Á and Zsigmond Zs 2009 *MNRAS* **398** 2199
Stern A and Cheng A 2002 *34th COSPAR Scientific Assembly* Houston, USA
Tholen D J, Buie M W, Grundy W M and Elliott G T 2008 *The Astr. J.* **135** 777
Weaver H A et al. 2005 *IAU Circular* 8625
Weaver H A et al. 2006 *Nature* **439** 943

Journal of Physics: Conference Series **218** (2010) 012021 doi:10.1088/1742-6596/218/1/012021

A young star's hectic months

Elza Elek[1] **and Mária Kun**[2]

[1] Eötvös University, H-1117 Budapest, Pázmány Péter sétány 1/A, Hungary
[2] Konkoly Observatory, H-1121 Budapest, Konkoly Thege út 15–17, Hungary

E-mail: eelza@szombat.konkoly.hu

Abstract. We present preliminary results of the optical photometric and spectroscopic monitoring of the young eruptive variable star PV Cephei, performed during its recent conspicuous fading in 2008–2009, and try to find the reasons of the observed variations.

1. Introduction

Young solar-type stars are highly variable across the whole electromagnetic spectrum. The main source of the variability are various types of interaction of the star with the surrounding accretion disk. Accretion disks evolve rapidly, giving rise to conspicuous variations of the central star: disk structures change significantly within 2–4 million years (e. g. Dahm & Carpenter 2009), and most primordial disks disappear by an age of 10 Myr (e.g. Sicilia-Aguilar et al. 2006). Variability studies offer a useful tool for probing the disk structure and evolution (e.g. Juhász et al. 2007). Causes of photometric and spectroscopic variability of young stars in the optical wavelength region include cool spots, induced by magnetic activity, hot spots, produced by infalling gas, and variable circumstellar extinction (Herbst et al. 1994, Scholz et al. 2009). Eruptive young stars (FUors an EXors), characterized by occasional, powerful brightening due to enhanced disk accretion, represent an especially interesting subclass of young variable stars. The target of our studies, PV Cephei, is a well known young eruptive star. It produced a large outburst in 1977–1978 when it brightened from 16th to 11th magnitude in the red (Cohen et al. 1981). This event ranked the star into the class of EXor type young stars, characterized by recurrent outbursts, lasting for several months. After the outburst in 1977–1978, no similar brightening of PV Cep has been recorded in the literature. The remarkable fading of the star between April and June 2008 in the near infrared was reported by Lorenzetti et al. (2009). We performed long-term optical photometric and spectroscopic monitoring observations of PV Cep in 2005–2009. Our goal is to better understand the nature of the observed variations. This paper presents our preliminary results.

2. Observations

Photometric observations were performed using the Schmidt and RCC telescopes of the Konkoly Observatory at Piszkéstető Mountain Station between 2005 and 2009 as part of a long-term observational project involving several other young stars. Spectroscopic observations were obtained at five epochs using the CAFOS instrument installed on the 2.2 m telescope of Calar Alto Observatory. The data reduction was carried out using standard IRAF routines. Detailed description of the data and results will be presented in a forthcoming paper.

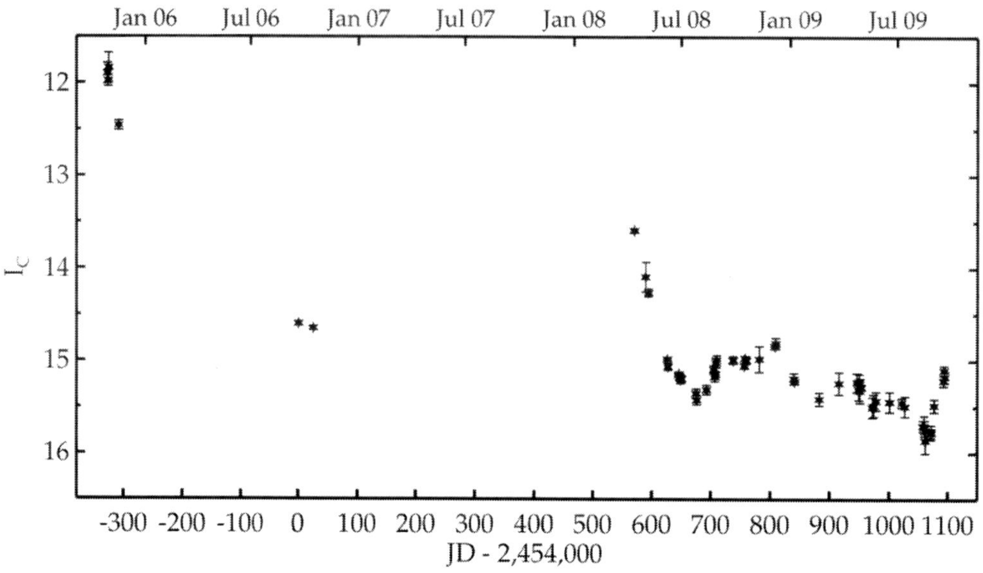

Figure 1. I_C-band light curve of PV Cephei between 2005 and 2009

3. Results

3.1. Light curve of PV Cep

Figure 1 shows the light curve of PV Cephei in the I_C band between October 2005 and September 2009. The sporadic data, available before April 2008, show that PV Cep was significantly brighter in 2005 than later. A steep decline of the light curve, similar to that reported by Lorenzetti et al. for the near infrared, started in April 2008. The light curve suggests that we probably observed the end of a large outburst of the star between April and June 2008, resulting from the significant drop of the disk accretion rate. The JHK two-color diagram (Fig. 10 of Lorenzetti et al. 2009), however, indicates that the fading might be caused by enhanced extinction along the line of sight. To clarify the nature of the fading and the variations during the present low-brightness state we examined the relations between the flux and color changes as well as the optical spectroscopic behavior of the star in the period covered by the light curve in Fig. 1.

3.2. Possible origin of variations

Cool spots as source of variability are obviously not relevant in the case of PV Cephei, because the light curve is not periodic, and the amplitude of variations is larger than it would be expected (cf. Herbst et al. 1994). Hot spots on the surface of accreting young stars arise where the infalling gas hits the photosphere. Changes in accretion rate or in the geometry of infall can lead irregular photometric changes on broad timescales. The amplitude of photometric variations caused by hot spots decreases toward longer wavelengths. If hot spots are responsible for the photometric variability, the star becomes redder when fades. We estimated the amplitudes of variations in the $R_C I_C JHK$ bands, caused by hot spots for several spot temperatures between 6500 and 12000 K and surface filling factors between 0.01 and 0.4. We assumed an effective temperature of $T_{\rm eff} = 5500$ K for PV Cep.

The left panel of Fig. 2 shows the measured R_C magnitudes against the $R_C - I_C$ color indices. Star symbols show the measurements before the fading, and dots show the low-brightness phase. The arrow indicates the displacement introduced by an extinction of $A_V = 1$ mag. As an example for the effect of hot spots, the length and direction of the dashed line shows how the star would

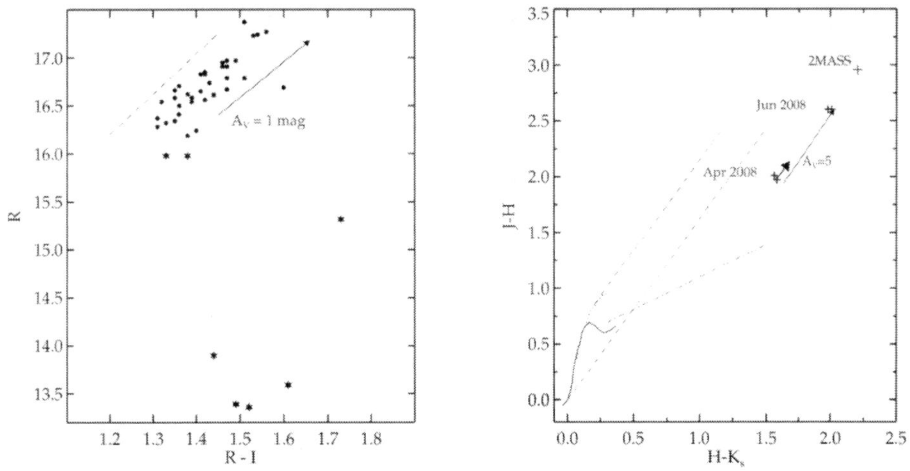

Figure 2. eft: R vs. $R-I$ diagram of PV Cep between Oct 2005 and Sep 2009. Star symbols show data obtained during the bright phase before June 2008. Right: Variation of the $J-H$ and $H-K$ color indices during the fading in 2008.

fade and redden if the hot spots (T_{spot} = 10000 K) covering 30% of its surface were removed. It can be seen that the variations during the dim phase can be explained by both extinction and disappearance of hot spots, whereas neither of these mechanisms work for the main fading. During the drop of the brighness the star became bluer, probably due to the higher proportion of the scattered light. The right panel displays the near-infrared two-color diagram, based on published data (2MASS, Lorenzetti et al. 2009). The color variation of the star between April and June 2008 can readily be explained by an extincion of $A_V = 5$ mag. The short arrow shows the effect of hot spot removal, assuming that in the bright phase 40% of the stellar surface was covered by spots of temperature of T_{spot} = 12000 K. Obviously, variable hot spots could produce much lower amplitude variations than observed.

According to the optical and near infrared observational results, the most likely reason of the photometric changes of PV Cep is apparently the variable extinction along the line of sight. The source of the extinction is probably the circumstellar dust, like in UX Orionis type stars, where orbiting dust clumps occasionally obscure the star. The observed light curve, however, markedly differs from the UXor type light curves (Herbst et al. 1994, Ismailov 2005). Both the amplitude and timescale of the fading, as well as the variations of near infrared color indices during the fading are remarkably similar to the behavior of the V1647 Ori at the end its outburst in October–December 2005 (Aspin & Reipurth 2009).

Spectroscopic observations of PV Cep at different epochs confirm that the accumulation of absorbing dust was accompanied by spectral variations indicative of variable accretion and wind. Figure 3 shows the time variation of the [OI], Hα, and [SII] lines during our observing period. For comparison, a bright-phase spectrum, obtained in 2004, is also shown.

4. Conclusion

The observed behaviour of PV Cep in 2008–2009 allow us to conclude that the star returned from an outburst into quiescence in spring 2008. Curiously, the beginning of the outburst remained undetected. According to Movsessian et al. (2008), PV Cep was bright in 2004. Unfortunately no photometric data are available for 2004. The bright phase continued in 2005, then the star faded between Nov 2005 and Sep 2006, and brightened again by April 2008. Similar large-amplitude variations were observed during the outburst in 1977–1978.

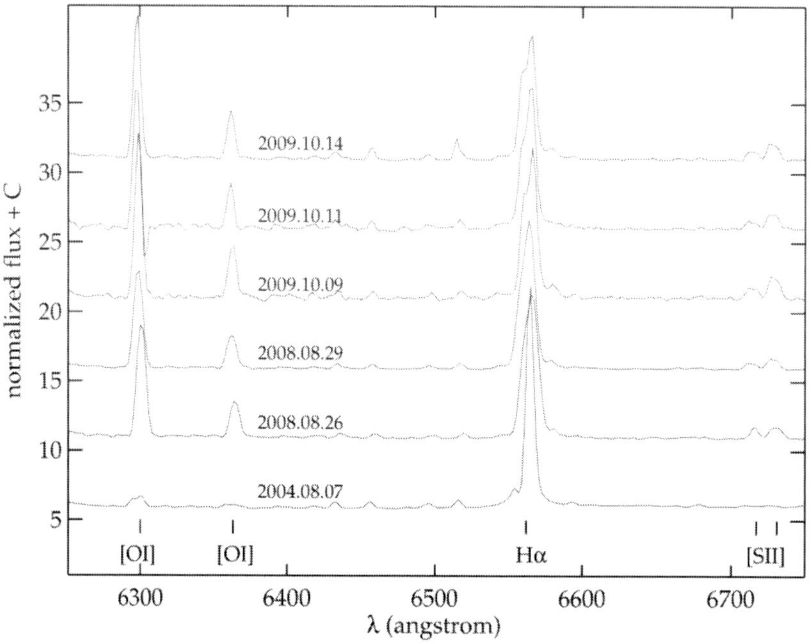

Figure 3. Five optical spectra of PV Cep in the red spectral region, obtained during the recent dim phase, and one obtained during the bright phase in 2004. The variable shape of the Hα line and the changing [SII] 6717/6731 flux ratio are indicative of the variable wind.

Acknowledgments

A significant part of the photometric observations have been performed by Attila Moór and Miklós Rácz. The spectroscopic observations were performed at the Centro Astronómico Hispano Alemán (CAHA) at Calar Alto, operated jointly by the Max-Planck-Institut für Astronomie and the Instituto de Astrofísica de Andalucía (CSIC). Our observations were supported by the OPTICON. OPTICON has received research funding from the European Community's Sixth Framework Programme under contract number RII3-CT-001566.

References

Aspin C, & Reipurth B 2009, *AJ* **138** 1137
Cohen M, Kuhi L V, Harlan E A, Spinrad H 1981, *ApJ* **245** 920
Dahm S E, & Carpenter J M 2009, *AJ* **137** 4024
Eiroa C, Oudmaijer R D et al. 2002, *A&A* **384** 1038
Herbst W, Herbst D K, & Grossman E J 1994, *AJ* **108** 1906
Ismailov N Z 2005, *Astron. Reports* **49** 309
Juhász A, Prusti T, Ábrahám P, Dullemond C P 2007, *MNRAS* **374** 1242
Lorenzettti D et al. 2009, *ApJ* **693** 1056
Movsessian T A, Magakian T Yu, Sargsyan D M, & Nikogossian E H 2008, *Astrophysics* **51** 387
Scholz A, Xu X, Jayawardhana R et al. 2009, *MNRAS* **873** 398
Siciila-Aguilar A, Hartmann L, Calvet N et al. 2006, *ApJ* **638** 897

5th Workshop of Young Researchers in Astronomy & Astrophysics IOP Publishing
Journal of Physics: Conference Series **218** (2010) 012022 doi:10.1088/1742-6596/218/1/012022

AKARI results on the Taurus-Auriga star forming region

Sarolta Zahorecz[1], L. Viktor Toth[1], Ueno Munetaka[2], Gabor Marton[1], Akiko Kawamura[3], Motohide Tamura[4]

[1] Department of Astronomy, Eötvös Loránd University, P.O. Box 32, H-1518 Budapest, Hungary

[2] Department of Earth Science and Astronomy, Graduate School of Arts and Sciences, The University of Tokyo, 3-8-1 Komaba, Meguro, Tokyo 153-8902, Japan

[3] Division of Particle and Astrophysical Sciences, Graduate School of Science, Nagoya University, Furo-cho, Chikusa-ku, Nagoya 464-8602, Japan

[4] National Astronomical Observatory of Japan, Mitaka, Tokyo 181-8588, Japan

E-mail: szahorecz@enkidu.elte.hu

Abstract. We present an analysis of AKARI Far-Infrared Surveyor, Spitzer Space Telescope Infrared Array Camera and Mid-Infrared Photometer for Spitzer data as well as various other photometric data of the Taurus-Auriga region. Our aim was to locate and describe young stellar objects (YSO), and to test the potential of AKARI FIS data in the classification of YSOs. We studied 32 YSOs, 7 of those had not been observed by the SST. We succesfully modeled 21 YSOs with the SED Fitting Tool of Robitaille. We confirmed the previous models in 20 cases, and corrected the physical parameters for 5 YSOs. This research is part of the AKARI Mission Program "Star Formation".

1. Introduction

Taurus-Auriga is one of the nearest and best-studied regions of low mass star formation (Kenyon et al. 1994), and contains 200 loosely clustered young stars. These stars are associated with long filaments and small clumps of dense gas (Jones & Herbig 1979, Gomez et al. 1993). There are deeply embedded sources in each small clump still accreting material from their own molecular cloud core (Beichman et al. 1986). Luminosities of most of these sources are similar to the Sun's and their estimated ages are from 10^5 to a few times 10^6 years (Myers et al. 1987, Strom et al. 1989, Kenyon et al. 1990, Gomez et al. 1992).

2. Observations

The main objective of the AKARI (Murakami et al. 2007) infrared astronomical mission is to carry out the All-Sky Survey in four photometric bands in the far-infrared wavelengths in 50−180 μm range with the Far-Infrared Surveyor (FIS; Kawada et al. 2007), and in two mid-infrared bands centered at 9 and 18 μm with the Infrared Camera (IRC; Onaka et al. 2007) with a sensitivity one order of magnitude better and resolution a few times higher than IRAS. AKARI is the second space mission for infrared astronomy in Japan. AKARI has been developed by members of JAXA/ISAS and collaborators.

2.1. *Far-Infrared Surveyor*
We used the data of Far-Infrared Surveyor (FIS). It provides four photometric bands between 50 and 180 μm, two broad bands and two narrow bands. The central wavelengths are: 65, 90, 140 and 160 μm. The band widths are about 21.7, 37.9, 52.4 and 34.1 μm.

2.2. *Bright Source Catalogue*
The Bright Source Catalogue (BSC) is the primary catalogue from the AKARI survey. It contains 63370 sources. 19085 sources have at least two good quality fluxes. There are only 3401 sources which have four good quality fluxes. In this catalogue, Version β-1 (Yamamura, 2008), the average position uncertainty is 8", and the estimated absolute flux uncertainty is 20-25 %. We compared the 2MASS positions with AKARI FIS BSC positions in the Taurus-Auriga region: we got an average 4.36" positional difference with a standard deviation of about 2".

3. Point sources in the Taurus-Auriga region
In the Taurus-Auriga region there are 173 AKARI FIS BSC point sources. There are 46 sources which have at least two good quality fluxes. We examined only these objects. With the use of 2MASS K images and VizieR catalogues, we identified the objects. There are 30 young stellar objects, 2 infrared sources, 11 galaxies. There are 3 objects which we could not identify clearly. In the figure 1 we can see the AKARI FIS BSC sources on the 13CO background (Dame et al. 2001). CO contour level is at 1.5 K km/s. In the left panel we show the point sources which have only 1 good quality flux, and in the right panel, the sources with at least two good quality fluxes.

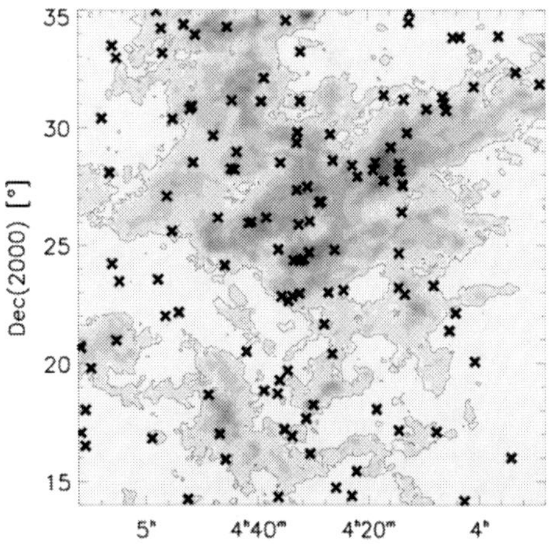

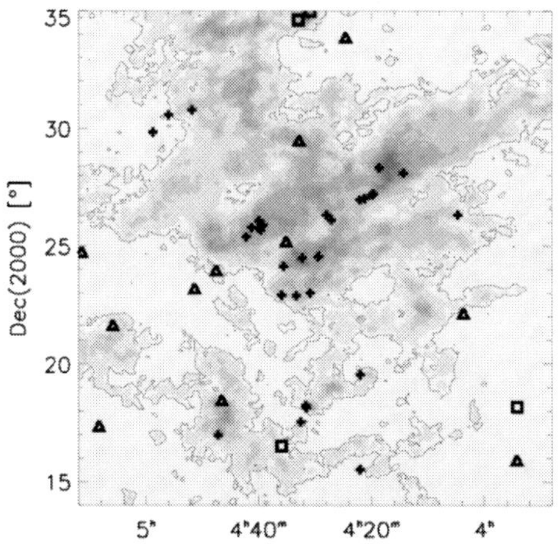

Figure 1. Left: AKARI FIS BSC point sources in the Taurus-Auriga region with 1 good quality flux. Right: Identified AKARI FIS BSC sources in the Taurus-Auriga region having at least two qood quality fluxes on the 13CO background (Dame et al. 2001). CO contour level is at 1.5 K km/s. Pluses are associated young stellar objects (30 objects), triangles indicate associated galaxies (11 objects) and squares are other objects (5 objects).

4. Models
We used the SED Fitting Tool of Robitaille (2006, 2007), which is publicly available on a dedicated web server. These model SEDs were computed using a Monte-Carlo radiation transfer

code, for 20071 different sets of physical parameters (e.g. stellar mass, stellar temperature, disk mass, etc...), and for ten viewing angles, resulting in a total of 200710 SEDs. It's a grid of radiation transfer models of axisymmetric young stellar objects, covering a wide range of stellar masses and evolutionary stages. The grid consists of 20000 YSO models, with spectral energy distributions (SEDs) and polarization spectra computed at 10 viewing angles for each model, resulting in a total of 200000 SEDs. Since none of the models have physical sizes larger than 100000 AU, the fluxes in the largest aperture correspond to the total fluxes.
We studied in detail only that point sources, which we could classified YSO or IR source. We try to model 32 objects (30 YSO and 2 IR sources). We could fit 21 point sources with the SED Fitting Tool. 13 of them was modeled previously by Robitaille et al. In these cases we use the data for their article and complete them with the AKARI FIS BSC fluxes. Our results are in good correlation with Robitaille's results. In the other cases we use the AKARI FIS BSC data and archive optical (VizieR) and infrared (SST IRAC and MIPS) data.

4.1. IRAS 04361+2547 and IRAS 04368+2557

IRAS 04368+2557 is the driving source of the molecular outflow in L1527. It is believed to be a single very young protostellar source. IRAS 04361+2547 is a typical Class I protostar (e.g. Tamura et al. 1991).
In figure 2 we can see the spectral energy distribution of IRAS 04361+2547 (left panel) and the SED of IRAS 04368+2557 (right panel). Data and errors are from Robitaille et al. (2006) and figure shows the new AKARI FIS BSC data as well. Upper limits are shown by arrows. Flux errors and FWHM of filters can be found, too. While getting the parameters we took the 10 best fit into account. These fits are shown by the grey zone.
In table 1 we can see the fitted parameters of modeled stars in the Taurus-Auriga region with 4 good quality fluxes.

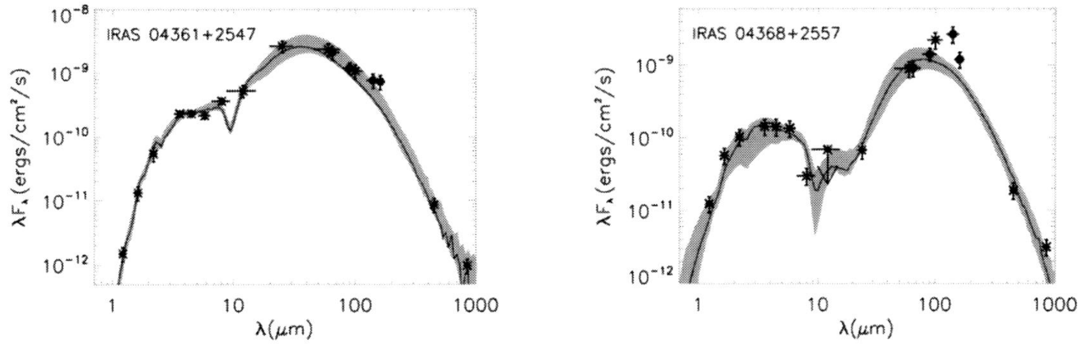

Figure 2. SED of IRAS 04361+2547 (left) and the SED of IRAS 04368+2557 (right). Asterisks show the archive data and points indicate the new AKARI FIS BSC data. Upper limits are shown by arrows. Flux errors and FWHM of filters can be found, too. While getting the parameters we took the 10 best fit into account. These fits are shown by the grey zone.

5. Results

We studied 32 YSOs in the Taurus-Auriga region, 7 of those had not been observed by the SST. We use new AKARI FIS BSC data and we succesfully modeled 21 YSOs with the SED Fitting Tool of Robitaille (2006, 2007). We confirmed the previous models in 20 cases, and corrected the physical parameters for 5 YSOs. This research is part of the AKARI Mission Program "Star Formation".

Journal of Physics: Conference Series **218** (2010) 012022 doi:10.1088/1742-6596/218/1/012022

Table 1. Fitted parameters of modeled stars in the Taurus-Auriga region with 4 good quality fluxes

Name	$t_\star$ [10^3yr]	$M_\star$ [$M_\odot$]	$R_\star$ [$R_\odot$]	$T_\star$ [K]	M_{disk} 10^{-4}[$M_\odot$]	Type
IRAS 04016+2610	$1.1 \pm^{4.1}_{0.0}$	$0.15 \pm^{0.09}_{0.05}$	$5.04 \pm^{0.19}_{1.28}$	$2847 \pm^{334}_{295}$	$154\pm^{0.0}_{95.6}$	Class I
IRAS 04169+2702	$193 \pm^{0}_{187}$	$0.43\pm^{0.00}_{0.30}$	$3.39\pm^{0.86}_{0.29}$	$3649 \pm^{0}_{923}$	$224\pm^{8}_{170}$	Class I
IRAS 04190+1924	$20.5\pm^{544.0}_{0.0}$	$0.70\pm^{4.43}_{0.00}$	$6.60\pm^{6.74}_{0.03}$	$3889\pm^{1521}_{0}$	$657\pm^{683}_{231}$	Class II
L1551 IRS 5	$6.3\pm^{32.1}_{5.2}$	$0.46\pm^{1.78}_{0.25}$	$6.00\pm^{6.45}_{1.39}$	$3651\pm^{668}_{581}$	$352\pm^{828}_{347}$	Class I
HL Tau	$1.3\pm^{65.2}_{0.2}$	$0.35\pm^{0.92}_{0.14}$	$5.78\pm^{2.07}_{0.92}$	$3454\pm^{744}_{393}$	$143\pm^{777}_{136}$	Class I
IRAS 04325+2402	$91.9\pm^{19.1}_{81.8}$	$0.20\pm^{0.03}_{0.10}$	$2.80\pm^{0.72}_{0.35}$	$3034\pm^{128}_{448}$	$1.04\pm^{8.11}_{0.789}$	Class I
IRAS 04361+2547	$1.2\pm^{141}_{0.0}$	$0.25\pm^{0.51}_{0.08}$	$5.23\pm^{0.70}_{1.15}$	$3225\pm^{787}_{290}$	$2.45\pm^{250}_{0.140}$	Class I
IRAS 04365+2535	$209\pm^{145}_{193}$	$0.87\pm^{0.34}_{0.69}$	$5.48\pm^{0.64}_{1.23}$	$4083\pm^{237}_{1080}$	$338\pm^{57.}_{203}$	Class I
IRAS 04368+2557	$237\pm^{0}_{127}$	$0.57\pm^{0.04}_{0.22}$	$4.10\pm^{0.27}_{1.09}$	$3855\pm^{37}_{374}$	$2.32\pm^{163}_{1.74}$	Class 0

Acknowledgments
I would like to thank Péter Ábrahám, Attila Moór, Nikolett Sipos for all their help and useful advices.
This research is based on observations with AKARI, a JAXA project with the participation of ESA. This work is based in part on observations made with the Spitzer Space Telescope, which is operated by the Jet Propulsion Laboratory, California Institute of Technology under a contract with NASA. This research has made use of the SIMBAD and VizieR data base operated at CDS, Strasbourg, France. This publication makes use of data products from the Two Micron All Sky Survey, which is a joint project of the University of Massachusetts and the Infrared Processing and Analysis Center/California Institute of Technology, funded by the National Aeronautics and Space Administration and the National Science Foundation.

References

Beichman, C. A. et al. 1986 *ApJ* **307** 337
Dame, T.M., Dap Hartmann and Thaddeus, P. 2001 *ApJ* **547** 792
Gomez, M. et al. 1992 *AJ* **104** 762
Gomez, M., Kenyon, S. J. and Hartmann, L. 1994 *AJ* **107** 1850
Jones, B. F. and Herbig, G. H. 1979 *AJ* **84** 1872
Kawada, M. et al. 2007 *PASJ* **59** S389
Kenyon, S. J., Hartmann, L. W., Strom, K. M. and Strom, S. E. 1990 *AJ* **99** 869
Kenyon, S. J., Gomez, M., Marzke, R. O. and Hartmann, L. 1994 *AJ* **108** 251
Murakami, H. et al. 2007 *PASJ* **59** S369
Myers, P. C., Fuller, G. A., Mathieu, R. D., Beichman, C. A., Benson, P. J., Schild, R. E. and Emerson, J. P. 1987 *ApJ* **319** 340
Onaka, T. et al. 2007 *PASJ* **59S** 401O
Robitaille, T. P., Whitney, B. A., Indebetouw, R., Wood, K. and Denzmore, P. 2006 *ApJS* **167** 256
Robitaille, T. P., Whitney, B. A., Indebetouw, R. amd Wood, K. 2007 *ApJS* **169** 328
Strom, K. M. et al. 1989 *AJ* **97** 1451S
Tamura, M. et al. 1991 *ApJ* **374** L25
Yamamura M. et al. 2008 AKARI/FIS All-Sky Survey Bright Source Catalogue Version β-1 – Release Note (Rev. 2) –

CO survey of ARCHEOPS cold cores

Erika Verebélyi[1], **Martin Miller**[2], **L. Viktor Tóth**[1], **Zoltán Makai**[2], **Gábor Marton**[1]

[1] Eötvös Loránd University, Budapest, Hungary
[2] I. Physikalisches Institut, Cologne, Germany

E-mail: everebelyi@astro.elte.hu

Abstract. Density and velocity distribution of 13 molecular clouds have been derived from multi-isotopic multi-line CO observations. These observations are investigations of the initial conditions for star formation in cold molecular cloud cores. The targets were selected from the ARCHEOPS point sources (Désert et al. 1987). The ARCHEOPS balloon born submm-mm telescope located cold interstellar clouds with $\sim$ 12 arcmin angular resolution (Désert et al. 1987, Benoit et al. 2002). A sub-sample of 13 clouds have been mapped in ^{12}CO(3-2), ^{13}CO(2-1) and C^{18}O(2-1). The observations have been carried out with the KOSMA-3m telescope (Kramer et al. 2000) in February, 2009. The measuring's pointing accuracy was around 10 arcsec. The calibration error was about 10-15%. The derived distribution of the gas column density indicated structured isolated molecular clouds with embedded cloud cores in 9 cases. Velocity gradient was found in 4 cases. The data set will be further used as input for Herschel target selection, observation and radiative transfer modelling.

1. Introduction

Dense region of the molecular clouds are the birthplaces of stars. They can be gravitational bounded cores or diffuse group of clumps. The local environment and interactions play a dominant role in the life of these clouds and affect the efficiency of star formation. An external influence changes the structure and shape of the parent molecular cloud. The line-shape and velocity-distribution inside the cloud provide us indirect information on the external influence. (Steinacker et al. 2005, de Vries et al. 2007, Horváth et al. 1995) It would be important to explore the interstellar medium, locate molecular cloud cores and derive their main parameters (temperature, and density inside the cloud) and examine the impact of external forces at barely known regions. Test object might be selected from a large area survey of cold galactic ISM, like the Archeops balloon-born experiment. Description of the instrument, the observational strategy and calibrations are presented F.-X. Désert et al. (2008). This is what we have done and this paper discusses the observation and analysis of CO measurement of cold galactic cores selected from the Archeops bright submillimetre point source catalog.

2. Target selection

A comparison of the Archeops bright point source-list to the Nagoya ^{13}CO-survey by Yonekura et al.(1997) and Dobashi et al. (1994) were completed. A sub-sample of 13 cold cores were chosen which have low dust temperature and were associeted with a ^{12}CO or ^{13}CO peak and preferably with extinction. (Kiss, 2009) The selected sources are: Arch103.97+10.93, ArchG106.88+05.29, ArchG108.11+02.75, ArchG108.80+01.40, ArchG110.26+02.46, ArchG115.83+03.97, ArchG115.92-01.62, ArchG117.33+03.14, ArchG120.67+04.02, ArchG121.76+00.22, ArchG124.54+02.09, ArchG127.69+02.65, ArchG128.93+04.35.

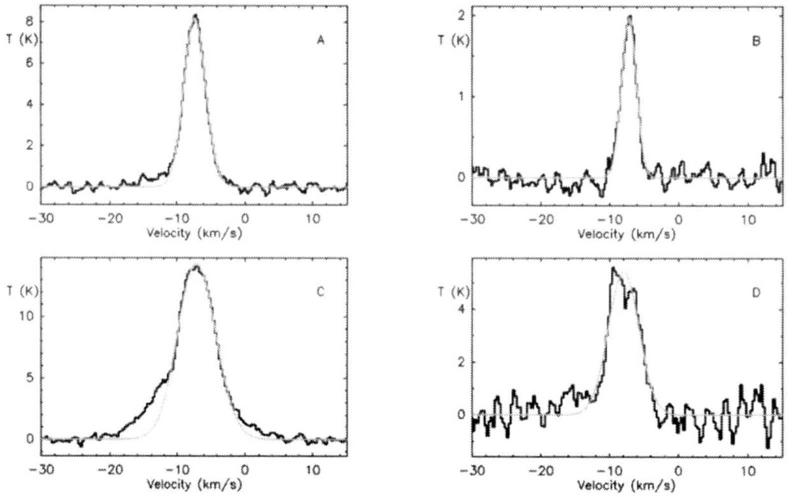

Figure 1. Calibrated spectra consist of the maximum intensity of the $^{13}CO(2\text{-}1)$ [A], $C^{18}O(2\text{-}1)$ [B], $^{12}CO(3\text{-}2)$ [C] observation. Black lines represent the measured spectra, and grey lines the fitted Gaussian curves. There are velocity in km/s on the horizontal axis compared to the LSR, and brightness temperature in K on the vertical axis. A typical $^{12}CO(3\text{-}2)$ line spectra from the "tail" is showed in [D].

Our analysis will be reviewed for one chosen cloud, the ArchG106.88+05.29 because there are the strongest ^{12}CO (3-2) and ^{13}CO (2-1) lines and we could detect strong $C^{18}O$ (2-1) lines, too.

3. Observations

Multiisotropic, multilevel CO-observations have been carried out with the KOSMA-3m (Kramer et al., 1998) telescope between 6 and 14, February in 2009. The telescope is located at Gornergrat in Switzerland, and is maintained by the I. Physikalisches Institut (Cologne, Germany) and Radioastronomisches Institut, University of Bonn. The observed transitions were $^{12}CO(3\text{-}2)$, $^{13}CO(2\text{-}1)$ and $C^{18}O(2\text{-}1)$ at 345.796, 220.399 and 219.560 GHz, respectively. The half power beamwidth of the telescope at these frequencies was 80, 120 and 120 arcsec. We used the OTF (on-the-fly) mode during the observation (Kramer et al. 2000). OFF positions were picked from the CO survey of Dame et.al. (Dame et al. 1987). The integration time per positions was 4 seconds, and the grid-size was 60 arcsec. At first the $^{12}CO(3\text{-}2)$ and the $^{13}CO(2\text{-}1)$ observation were done simultaneously, than having chosen the maximum position the $C^{18}O(2\text{-}1)$ measurement was done to a smaller area, with a longer integration time. Our pointing sources were: Venus and Saturn. Our calibration sources were: DR21, Orion KL and W51A. The position error was about 10 arcsec and the calibration error was 10-15 %.
After the observation the necessary calibration steps were done on the spot using the GILDAS data analysis package following the (Kramer et al. 2000).

4. Analysis

Calibrated sample-spectra are shown in the Figure 1. The lines from the ^{13}CO (2-1) and $C^{18}O$ (2-1) observations are mainly Gaussian (just like the sample-spectra) but in the core the ^{12}CO (3-2) line spectra have a prominent wing in the blue side which suggests a presence of outflow from an embedded YSO (see Fig. 1C). However, if we look to the tail and not to the core of the cloud we can get a different characteristic. Figure 2. shows the distribution of the three different CO-isotopomers. To follow the way of the definition of the core-size written in Wu

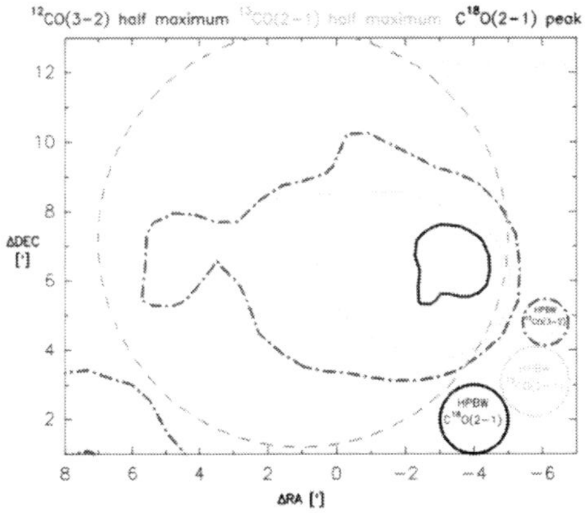

Figure 2. Complex picture from the cloud: the different lines represent the half maximum or the peak intensity-contour of the three different transition. There are offset-coordinates in arcmin units on both axis. The big dashed round shows the resolution of Archeops.

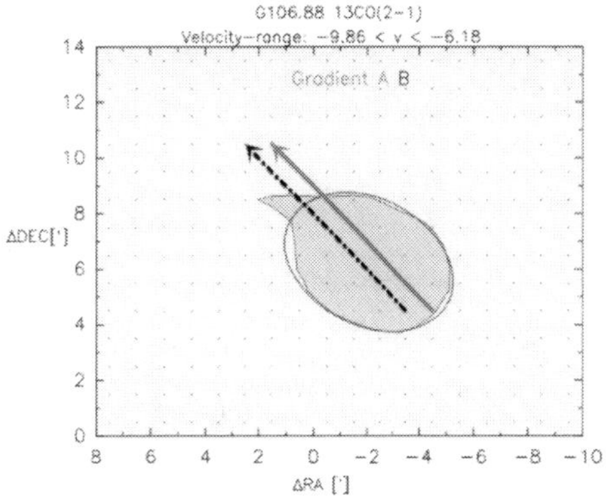

Figure 3. It shows the half maximum contour of the line area from the ^{13}CO(2-1) measurement with the fitted ellipse, and the two axis where we looked for the velocity and linewidth gradient. The small dots in the background indicate the grid of the observation.

et al. (2000), an ellipse was fitted to the half maximum contour of the ^{13}CO-intensity (Figure 3.) In this case the core is compact and spherical. After the basic calibration steps we can determine the brightness temperature from the spectra. Following the description of Wilson et al. (2008), the excitation temperature, the optical depth and the column densities are countable. The ^{12}CO excitation temperature of Arch106.88 is 21 K, and varied between 5 and 20 K for the other clouds of the sample. The column density to the ^{13}CO molecule has been also derived and we estimated the column density of the H_2-molecule as well. The ratio between H_2 and ^{13}CO is $N(H_2)=7{,}89\times10^5\ N(^{13}CO)$ taken from Beuther et al.(2000). All of these parameters are represented in the Table 1.

Arch106.88 has a head-tail shape so we studied the velocity-gradient in its axis of symmetry. These axis were shown in the Figure 2. We found a very weak velocity gradient in the ^{12}CO-lines, but in the ^{13}CO line the effect is not detectable. Figure 4. shows the line velocities along the two axis, from the $[-3'.5, 4'.5]$ to grad A, and from the $[-4'.5, 4'.5]$ offset coordinate (grad B) as an initial position. The distribution of linewidth (FWHM) was also checked and is shown in Fig. 4.

5. Results

- We detected ^{12}CO (3-2) and ^{13}CO (2-1) in all 13 selected cold Archeops clouds.
- $C^{18}O$ (2-1) was also detected in five of the clouds.
- Clouds structure and density distribution were obtained. We found a compact cloud core in 9 cases, and a diffuse structure in 4 cases.
- The sizes of the cores were estimated.
- The excitation temperature and the column densities of H_2 molecule were determined in the core position of each cloud.
- Velocity gradient was found in 4 cases.

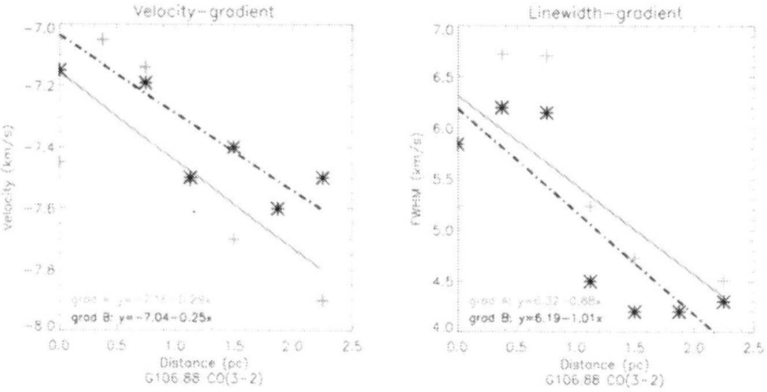

Figure 4. The velocity (left) and linewidth-gradient (right) along the two axis from the ^{12}CO-observation. There are distances on the vertical axis in the unit of pc, and there is a velocity and a linewidth on the horizontal axis. Both units are km/s. The equation of the fitted line are showed on the left corner of the diagram.

Name	Offset coordinate	a	b/a	PA [deg]	T_{ex} [K]	$N(H_2)$ [$\times 10^{21}$cm^{-2}]	Distance [pc]
ArchG106.88+05.29	-2',1 : 6'.3	3',3	0,70	25	21,0	11,8	910

Table 1. The name of the source (1), central position of the ellipse in offset coordinate (2), the major axis of the fitted ellipse (3), the ratio of the minor and major axis (4), the position angle (the different from east) (5), the exciting temperature of the ^{12}CO (6), the column density of H_2 (7), and the estimated distance of the cloud from (Yonekura et al. 1997)(8).

The data will be further used as input for Herschel target selection, observation and radiative transfer modelling.

5.1. Acknowledgments
The KOSMA 3 m radiotelescope at Gornergrat-Sfid Observatory is operated by the University of Cologne and supported by special funding from the Land NRW. The Observatory is administered by the Internationale Stiftung Hochalpine Forschungsstationen Jungfraujoch und Gornergrat, Bern. This research was partly fund by OTKA K62 304.

References
Benoit A, Ade P, Amblard A, Ansari R, Aubourg E, Bartlett J, Bernard J-P, Bhatia R S, Blanchard A, et al 2002 *APh* **17** 101
Beuther H, Kramer C, Deiss B and Stutzki J 2000 *A&A* **362** 1109
Désert F-X, Macas-Prez J F, Mayet F, Giardino G, Renault C, Aumont J, Benoit A, Bernard J-Ph, Ponthieu N, et al. 2008 *A&A* **481** 411
Dobashi K, Bernard J-Ph, Yonekura Y and Fukui Y 1994 *ApJS* **95** 419
Horváth A and Tóth L V 1995 *Ap&SS* **233** 169
Kiss Z., private communication
Kramer C, Beuther H, Simon R, Stutzki J and Winnewisser G 2000 *ASPC* **217** 194
Steinacker J, Bacmann A, Henning Th, Klessen R and Stickel M 2005 *A&A* **434** 167
de Vries C H, Narayanan G and Snell R L 2007 *IAUS* **237** 409
Wilson T L, 2008 The Tools of Radioastronomy
Wu Y, Yan H, Wang J, Wu J, Zhao Y, Lei C, Sun J and Wang L 2000 *ASPC* **217** 96
Yonekura Y, Dobashi K, Mizuno A, Ogawa H and Fukui Y 1997 *ApJS* **110** 21

5th Workshop of Young Researchers in Astronomy & Astrophysics IOP Publishing
Journal of Physics: Conference Series **218** (2010) 012024 doi:10.1088/1742-6596/218/1/012024

Preliminary analysis of the AKARI FIS BSC

Gábor Marton[1], L. Viktor Tóth[1], Munetaka Ueno [2], Sarolta Zahorecz[1], Akiko Kawamura[3], Motohide Tamura[4]

[1] Department of Astronomy, Eötvös Loránd University, P.O. Box 32, H-1518 Budapest, Hungary

[2] Department of Earth Science and Astronomy, Graduate School of Arts and Sciences, The University of Tokyo, 3-8-1 Komaba, Meguro, Tokyo 153-8902, Japan

[3] Division of Particle and Astrophysical Sciences, Graduate School of Science, Nagoya University, Furo-cho, Chikusa-ku, Nagoya 464-8602, Japan

[4] National Astronomical Observatory of Japan, Mitaka, Tokyo 181-8588, Japan

E-mail: `g.marton@astro.elte.hu`

Abstract. We present a preliminary analysis of AKARI Far-Infrared Surveyor Bright Source Catalog. The AKARI mission collected data on 63370 bright IR sources. Galactic spatial distribution of different flux quality types was studied. The surface density shows correlation with already known places of IR bright galactic regions. Difference between the spatial distribution of different flux quality objects is not notable. This research is part of the AKARI Mission Program "Star Formation".

1. Introduction

The AKARI astronomical satellite onboard with the Far-Infrared Surveyor accomplished a full-sky survey. The survey is similar to former IR all sky surveys like 2MASS (Skrutskie et al., 1995) and IRAS Sky Survey (Beichmann, 1988) but with a sensitivity one order of magnitude better and resolution a few times higher than IRAS (Yamamura et al., 2008). Galactic IR-bright sources are mainly late type stars, circumstellar disks and clouds of interstellar dust and gas. Extragalactic objects like LMC and SMC may also have strong IR emission. Full-sky surveys are useful tools for unbiased studies of various types of objects. Our aim is to classify different types of objects with a special attention to YSOs. We investigate the galactic distribution of AKARI point sources and compare it to the interstellar CO distribution (Dame et al., 2001), a good tracer of cold ISM.

2. Observations

The main objective of the AKARI (Murakami et al. 2007) infrared astronomical mission was to carry out the All-Sky Survey with the Far-Infrared Surveyor (FIS; Kawada et al. 2007), and with the Infrared Camera (IRC; Onaka et al. 2007). AKARI is the second space mission for infrared astronomy in Japan. AKARI has been developed by members of JAXA/ISAS and collaborators. We used the data of Far-Infrared Surveyor (FIS). It provides four photometric bands between 50 and 180 μm, two broad bands and two narrow bands. The N60 and WIDE-S detectors are Monolithic Ge:Ga detectors while the WIDE-L and N160 detectors are Stressed Ge:Ga detectors (Yamamura et al., 2008). For more details see Table 1.

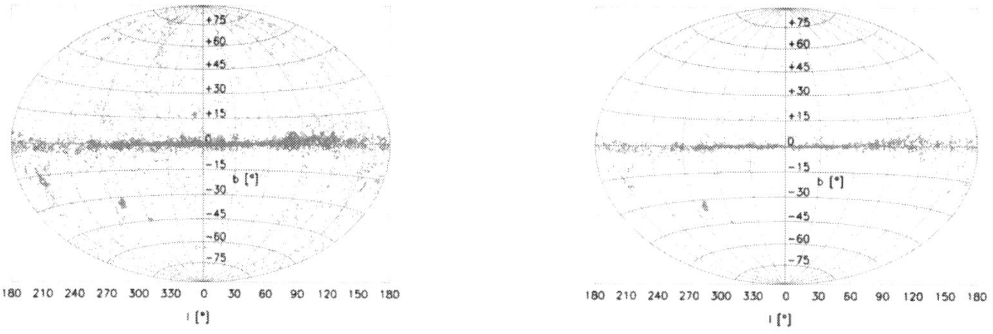

Figure 1. Galactic distribution of AKARI FIS BSC point sources in equal-area Aitoff-projection. Left: point sources with quality 3 WIDE-S and WIDE-L fluxes. Right: point sources with only quality 3 fluxes.

3. AKARI FIS BSC

The AKARI FIS Bright Source Catalogue (BSC) is currently the primary catalogue of the AKARI survey. It is also referred to as "AKARI point source catalogue". The β-1 version contains 63370 point sources. The average position uncertainty is 8", and the estimated absolute flux uncertainty is 20-25% (Yamamura ey al., 2008). The catalogue lists position, flux densities, flux qualities, number of scans and number of readouts for each sources. The calibrated fluxes measured by the detectors have 3 flux quality classes. These classes are based on the flux uncertainties. The highest value of 3 means the best quality. The evidence of the data is described with LogEvidence value. The value 3 means that LogEvidence ≥ 10.0. If LogEvidence < 10.0 than the value of the flux quality is 2. Flux quality 1 means that the data is not reliable. For future classification one should use the most reliable data where flux quality is 3. All of the point sources have at least one flux quality which equals 3, in 11905 cases at least two, in 3758 cases at least 3 and in 3401 cases all of the measured fluxes have the best quality.

4. Galactic distribution of AKARI point sources

4.1. Distribution as a function of galactic latitude

The galactic distribution of AKARI point sources is shown in Fig. 2, the histogram was made with 10° wide bins. The distribution shows a strong maximum in the direction of the galactic midplane. The width at half-maximum is 4°. There is a local maximum at $-30° < |b| < -40°$ on both panels, that is at the latitude of the LMC. The number of the sources where the WIDE-S and WIDE-L flux qualities are the best (hereafter *33* sources) has a powerful maximum in the direction of the galactic plane (Fig. 2).

4.2. Distribution as a function of galactic longitude

The distribution of the sources as a function of galactic longitude is shown in Fig. 3. The *33* sources' distribution has 2 minima. First is at the galactic anticentre $120° < |l| < 200°$ region inspite of the many YSOs seen at the Tau-Aur-Per complex. The other minimum is at $220° < |l| < 260°$, where also the amount of the interstellar medium is low, so YSOs cannot form in a large number. Similar distribution can be seen for the 3333 sources figure (right panel of Fig. 3).

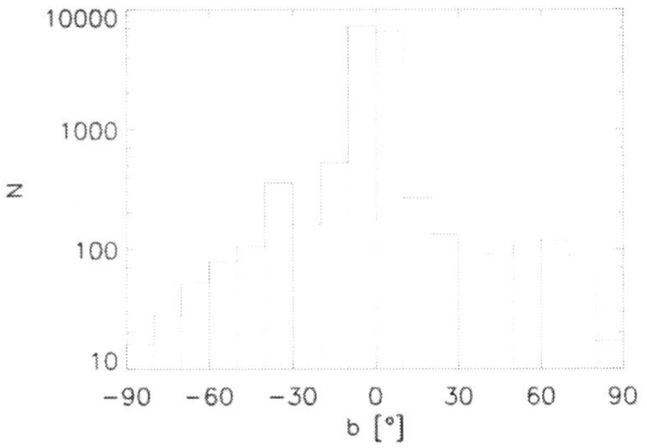

Figure 2. Histogram of the AKARI FIS BSC sources - distribution as a function of galactic latitude. Left: sources with quality 3 WIDE-S and WIDE-L fluxes. Right: sources having only quality 3 fluxes.

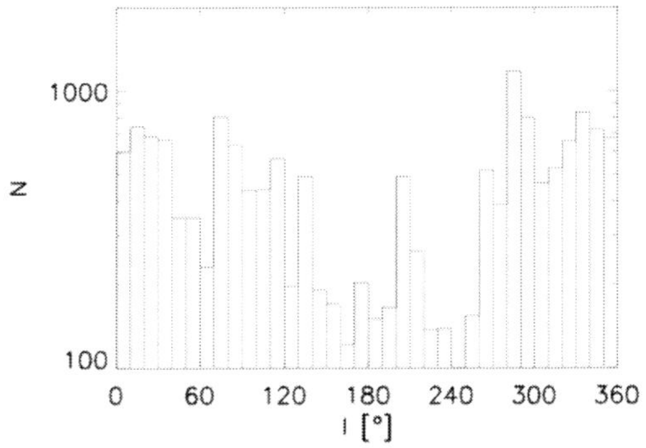

Figure 3. Histogram of the AKARI FIS BSC sources - distribution as a function of galactic longitude. Left: sources with quality 3 WIDE-S and WIDE-L fluxes. Right: sources having only good quality (3333) fluxes.

5. Discussion

The IR bright sources are mainly young stars or galaxies. The infrared excess of YSOs is caused by the pre-main sequence state, where the observed temperature is lower than in MS stars. The surrounding dust and gas also radiates on these wavelengths and the emitted light is also scattered on small particles. Extragalactic sources can have a notable amount of dust and cool gas, too, while their distance and the intergalactic matter causes a similar scattering effect. Stars are born where the quantity of gas is enough for star formation. These conditions are mainly available in the galactic disk, near the galactic plane (Dame et al., 2001). We note that 81.04% of the sources are between -5° and 5° galactic latitudes, 12.34% of the sources are at medium galactic latitude (5°$< |b| <$ 30°) and only 6.62% are at high $|b|$ values. See Fig. 1 and Fig. 2. The galactic distribution of the AKARI FIS BSC point sources shows strong correlation with the galactic distribution of CO (Dame et al., 2001). Looking at the distribution as a function of the galactic latitude both the point source distribution and the CO distribution has a maximum at $-10° < b < 10°$. The distribution of the AKARI FIS BSC point sources and the CO are also similar along galactic longitudes. One can see plateau at the inner Galaxy ($0° < l < 60°$ and $300° < l < 360°$) and a local minimum at $220° < l < 260°$ on Fig. 3 and Fig. 4.

Table 1. Main parameters of the AKARI FIS filters*

Band	N60	WIDE-S	WIDE-L	N160
Wavelength range (μm)	50-80	60-110	110-180	140-180
Central wavelength (μm)	65	90	140	160
Band width (μm)	21.7	37.9	52.4	34.1

* Yamamura et al., 2008

Acknowledgments
This research is based on observations with AKARI, a JAXA project with the participation of ESA. This research has made use of the SIMBAD and VizieR data base operated at CDS, Strasbourg, France.

References

Beichmann C A 1988 *LNP* **297** 148B
Dame, T M, Dap Hartmann and Thaddeus P 2001 *ApJ* **547** 792
Kawada M et al. 2007 *PASJ* **59** S389
Murakami H et al. 2007 *PASJ* **59** S369
Onaka T et al. 2007 *PASJ* **59S** 401O
Skrutskie M F et al. 1995 *AAS* **187** 7507S
Yamamura M et al. 2008 AKARI/FIS All-Sky Survey Bright Source Catalogue Version β-1 – Release Note (Rev. 2) –

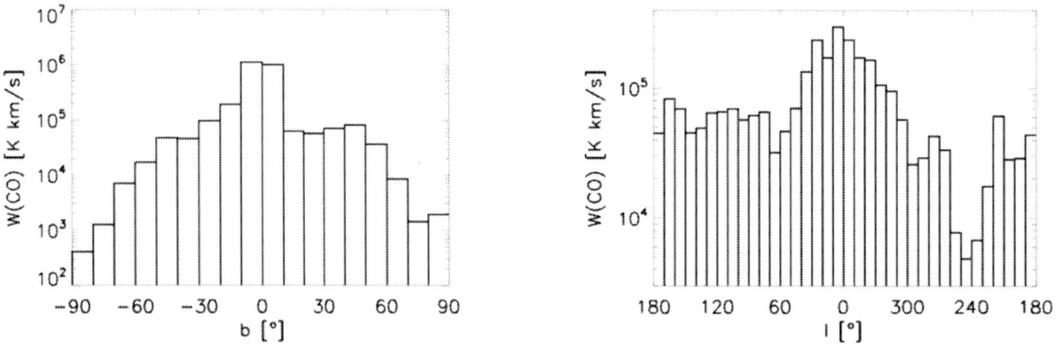

Figure 4. Histogram of the Whole-Galaxy velocity integrated CO map (Dame et al., 2001). Left: galactic latitude dependence of the velocity integrated CO line area. Right: galactic longitude dependence of the velocity integrated CO line area.

Effects of [Fe/H] on the properties of Galactic classical Cepheids

Péter Klagyivik and László Szabados

Konkoly Observatory, H-1525 Budapest XII, P.O. Box 67, Hungary

E-mail: klagyi@konkoly.hu, szabados@konkoly.hu

Abstract. Correlations have been searched for between the atmospheric [Fe/H] value and the observed amplitudes of classical Cepheids. Model calculations show that fundamental mode pulsation is unstable for solar like chemical composition between 8 and 10 days, but observations contradict the models. The ratio of the radial velocity and photometric amplitudes, $A_{V_{\rm RAD}}/A_B$, shows a mild metallicity dependence.

1. Introduction

Classical Cepheids are radially pulsating supergiant stars. Their pulsation is a free radial oscillation of the outer layers of the star. The pulsation is maintained by the κ-mechanism and is stable in a narrow zone, in the so-called instability strip in the Hertzsprung-Russell diagram. The radial oscillation and the instability strip result in many relationships between the properties of Cepheids. The most important is the period-luminosity (P-L) relationship.

Motivated by the fact that the dependence of the pulsation amplitude on $\log P$ is neither linear, nor unique, we revised the period-amplitude (P-A) relationship of Galactic classical Cepheids for the U, B, V, R_C, I_C photometric bands and the pulsational radial velocity variations (Klagyivik & Szabados 2009).

The different patterns of the P-A diagrams based on Galactic and Magellanic Clouds Cepheids show that the metallicity is a non-negligible factor of the pulsation amplitudes, first mentioned by van Genderen (1978). The role played by the metallicity in adjusting the P-L relationship has been frequently discussed from both observational and theoretical points of view (see the recent summary by Romaniello et al. 2008), though the results are controversial. Using the homogeneous OGLE data base, Paczyński & Pindor (2000) found that the higher the metal content of the host galaxy, the larger the photometric amplitude.

Here we present the results on the period dependence of [Fe/H] and the metallicity dependence of the amplitude ratio $A_{V_{\rm RAD}}/A_B$, i.e., the ratio of the peak-to-peak amplitudes of the radial velocity variations over the brightness variations in the Johnson B band.

We used a data base of observational data of 369 Galactic classical Cepheids. This catalogue contains the photometric and radial velocity amplitudes, some parameters derived from the amplitudes and the spectroscopically determined [Fe/H] ratio for 179 stars. For details see Klagyivik & Szabados (2009) where it is also shown that there exists a dichotomy separating short- and long-period Cepheids and the limiting period value is 10.47 days.

2. Discussion

2.1. Period dependence of [Fe/H]

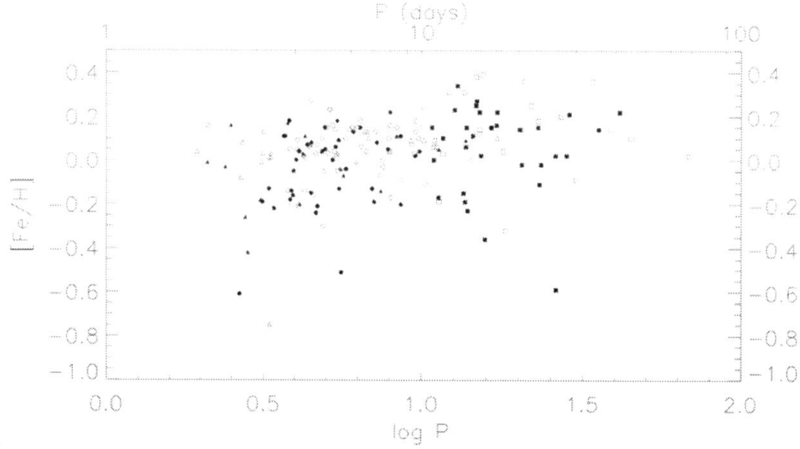

Figure 1. Period dependence of [Fe/H]. Circles represent fundamental mode Cepheids with short pulsation period (P < 10.47 days), squares mean long period stars (P > 10.47 days), while triangles correspond to first overtone Cepheids. Filled symbols represent solitary stars, while Cepheids denoted with open symbols have known companions.

According to the P-L relationship, long-period Cepheids are more luminous than their shorter period counterparts, and being more massive, they evolve rapidly. As a consequence, longer period Cepheids are also younger, and owing to the gradual enrichment of the heavy elements in the interstellar matter, the metallicity of the younger stellar population is larger. This trend is clearly seen in Fig. 1.

Buchler et al. (1997) found that the fundamental mode limit cycles are unstable in the period range 8-10 days for Cepheids with metallicity between $Z = 0.014$ and 0.035. This metallicity range corresponds to the −0.154 - 0.244 interval of [Fe/H] using the conversion relation $\log Z = \text{[Fe/H]} - 1.7$ (Caputo et al. 2001). The corresponding stars pulsate in the first overtone with a period of $P_1 \approx 0.7P_0$, causing a relative excess of first overtone Cepheids in the period interval of 5.6-7.0 days. Magellanic Clouds are metal deficient as compared with our Galaxy, therefore fundamental mode pulsation is allowed there in the 8-10 days period range. In spite of the model calculations the box that corresponds to the models in Fig. 2 is not empty: observations show that fundamental mode pulsators do exist in our Galaxy in this period range. The error of the individual [Fe/H] values is about 0.05-0.10 dex, but this fact may not fully explain the discrepancy between models and observations.

2.2. Metallicity dependence of the amplitude ratio q

Importance of the specific amplitude ratio $A_{V_{\rm RAD}}/A_B$ – referred to as the q parameter – was discussed by Klagyivik & Szabados (2009). This ratio is sensitive to the binarity, because companions reduce the observable photometric amplitudes, and an unrecognized orbital motion superimposed on pulsational changes results in an increased $A_{V_{\rm RAD}}$. Therefore in the case of testing the metallicity dependence of q we have to exclude binary Cepheids.

The relationship between q and metallicity is presented in Fig. 3. The symbols are the same as in Fig. 1. Linear least squares fits have been applied to each group of Cepheids separately and

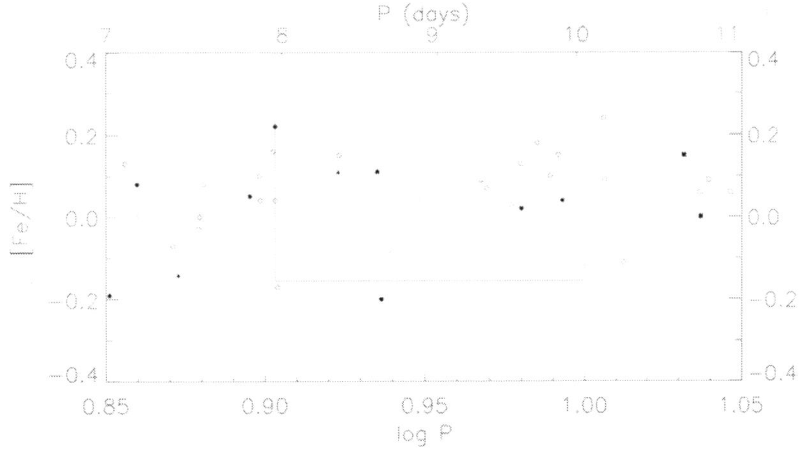

Figure 2. Distribution of Cepheid [Fe/H] values near the unstable zone of Buchler et al. (1997). Symbols are the same as in Fig. 1.

to all fundamental mode pulsators as well. The results are summarized in Table 1. Coefficients a and b correspond to the following formula:

$$q = a + b \times [\mathrm{Fe/H}] \tag{1}$$

Table 1. Coefficients of the linear fit to the q vs. [Fe/H] relationship (Eq.1)

Sample	a	σ_a	b	σ_b	N
$\log P < 1.02$	34.68	0.83	−14.09	7.34	23
$\log P > 1.02$	31.65	0.96	−3.53	5.54	23
All fundamental mode Cepheids	33.36	0.65	−9.45	4.49	46
First overtone Cepheids	35.28	1.01	−16.80	8.91	15

All types of Cepheids follow the same trend, the higher the metallicity, the lower the $A_{V_{\mathrm{RAD}}}/A_B$ ratio. These correlations are not very strong (2σ).

3. Conclusion

We searched for metallicity dependences of several parameters related to the pulsation of Cepheids. The correlations we have found are not very strong, but clearly demonstrate that the effect of the chemical composition on the pulsation is non-negligible. The individual [Fe/H] values have an error of 0.05-0.10, so a more precise formulation of these relations is mainly hindered by uncertainties of the abundance determination.

Acknowledgments

This research was supported by the European Space Agency (ESA) and the Hungarian Space Office via the PECS programme (contract No. 98090).

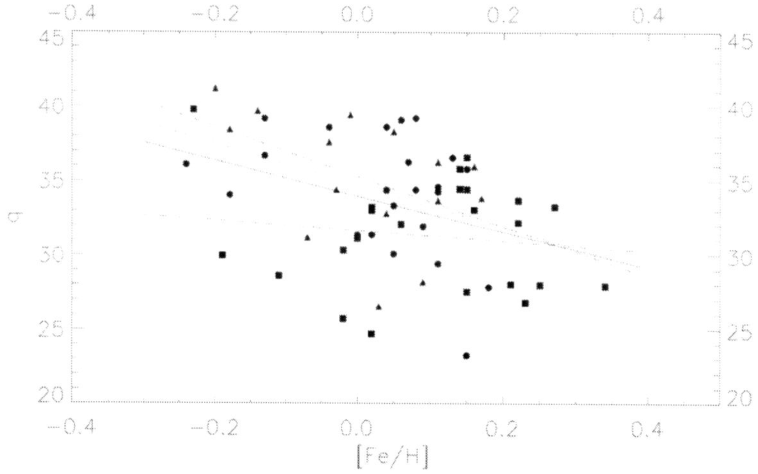

Figure 3. Metallicity dependence of the amplitude ratio q. The solid line corresponds to the fit based on all fundamental mode Cepheids, while dotted, dashed and dash-dotted lines represent the fit to short period, long period and first overtone Cepheids, respectively.

References

Buchler J R, Goupil M J and Piciullo R 1997 *Astrophys. Journ.* **491** L99

Caputo F, Marconi M, Musella I and Pont F 2001 *Astron. and Astrophys.* **372** 544

van Genderen A M 1978 *Astron. and Astrophys.* **65** 147

Klagyivik P and Szabados L 2009 *Astron. and Astrophys.* **504** 959

Paczyński B and Pindor B 2000 *Astrophys. Journ.* **533** L103

Romaniello M, Primas F, Mottini M, Pedicelli S, Lemasle B, Bono G, François P, Groenewegen M A T and Laney C D 2008 *Astron. and Astrophys.* **488** 731

5th Workshop of Young Researchers in Astronomy & Astrophysics IOP Publishing
Journal of Physics: Conference Series **218** (2010) 012026 doi:10.1088/1742-6596/218/1/012026

Automated data analysis – algorithms for Gaia Cepheids

Zoltán Tamás Kiss, László Szabados
Konkoly Observatory of the Hungarian Academy of Sciences,
Budapest 1525, P.O. Box 67, Hungary

E-mail: kissz@konkoly.hu

Abstract. Gaia space astrometric mission, to be launched in 2012, is expected to provide a bulk of data observing a thousand million objects within its five year time-span. Due to its scanning law the all sky will be covered, and all the available objects will be observed repeatedly with a moderate time sampling, therefore Gaia will obtain a large amount of photometric data that carry valuable information on a wide range of variability phenomena.

The Gaia group of the Konkoly Observatory works on the preparation of automated characterisation algorithms of pre-classified Cepheid type sources within the Variability Processing group of Gaia.

Gaia is expected to provide observational data on all available objects with a number of data points gradually increasing during the mission. Based on the scanning law a moderate number of data points per source are expected. That, though makes the processing necessarily to be fine tuned, is sufficient for the characterisation including determination of period(s), Fourier parameters, possible period changes, and hints on binarity. Therefore we use for testing public and unpublished data sources that are comparable to those expected from Gaia in term of sampling combined with simulated Gaia measurements. Depending on the phase coverage ground based supplementary observations may be needed to improve the accuracy of the results.

The final product of this work will be a contribution to the Gaia database, that will be compiled primarily based on the Gaia on-board measurements, and probably include a number of newly discovered Cepheids. This work may lead to an improved knowledge of the nature of these objects, that are fundamental to approach a more precise cosmic distance scale.

1. Introduction

1.1. Gaia overview

Gaia is designed to continue and extend the efforts made by the Hipparcos, the former astrometric space mission of the ESA, to draw the three dimensional picture of our astronomical environment. Gaia measures all objects detectable down to 20th mag with unprecedented accuracy, therefore the expected number of objects to be detected is around 10^9. The expected photometric and astrometric uncertainty is 0.05 mag at 20th mag and 24 μas at 15th mag. However this is not the only substantial improvement with respect to the Hipparcos. To determine all the kinematical properties of the observed objects, not just the accurate spatial location but also the complete velocity vector, Gaia measures radial velocities of the target objects performing spectroscopic measurements. Measurements are repeated several times for all objects during the mission in order to reveal possible temporal changes in the observable properties of astronomical objects. Scheduled for launch in 2012, Gaia will perform its measurements in the L2 Lagrange point of the Sun-Earth system rotating with a period of 6 hours around its spin axis that points 50 degrees from the direction of the Sun with a precession period of 70 days. The optical system has two distinct lines of sights with 106 degrees separation, both are perpendicular to the rotation axis. Due to this geometry and rotation each object crosses its fields of view 70 times on average during the 5 year expected lifetime of the Gaia. This number of observations per object allows us to study the variability of the objects detected. Within the frame of the Coordination Unit 7 (CU7) of the Gaia Data Processing and Analysis Consortium (DPAC) the Gaia group of the Konkoly Observatory is responsible for the development and testing of software related to data on Cepheid variables to be observed by Gaia.

1.2. Gaia Cepheid observations

Due to the fact that all detected objects will be measured several dozen times, Gaia allows us to investigate the brightness variations of the objects observed. Though the expected number of data points per objects is relatively small, 70 on average, in combination with the unprecedented photometric accuracy and number of detected objects it is still enough for accurate characterisation of light variations. The applied scanning law provides appropriate time sampling, leading to a good phase coverage, with the exception of those objects that have periods nearly in resonance with the Gaia's rotational or precession period.

The procedures being developed by the Konkoly Gaia group are meant to be used within the Gaia data processing pipeline. The input data are the pre-processed observations organised into photometric time series. During the pre-procession a period search and a classification are to be done. Our procedures work only with data of objects that were found to be variable and classified as Cepheids based on the data available in the earlier stages of the pipeline by procedures developed by the DPAC Coordination Units 1-6. Though the number of objects to be processed is not so huge as the total number of objects detected by Gaia, because of the possible misclassification, still much larger than the expected final number of Cepheids to be found by Gaia. The main issue, however, is the diversity of the possible input data. The resulting accuracy may be sensitive to the time sampling and periods of light variations as well as the possible variations in both period and light curve shape that are often unpredictable even for the known variables. The results to be obtained using algorithms that have to be applied as part of a pipeline are quite sensitive to the way how these effects are taken into account.

We perform the coding of the developed algorithms to Java language to make the source code fit to the Gaia framework, and make it applicable for automated operation in processing the future Gaia data, and implement the results in the Gaia catalogue.

2. Methods and input data

2.1. Methods

The procedures being developed are meant to determine the period or periods, calculate the coefficients of the Fourier decomposition of the light curves for the significant harmonics in all available colours. At first a Fourier decomposition is performed to find out the rough values of periods. In case of a single period it is refined applying a least squares fit to the light curve using a function in the following form:

$$m(t) = C + A_0 sin(2\pi(ft + \Phi_0)), \quad (1)$$

where t is the time in days with respect to an arbitrary initial epoch; m is the magnitude for the given t; C is the mean magnitude; A_0 is the amplitude for the main frequency f; and Φ_0 is the initial phase at $t = 0$. Phase values are ranging from 0 to 1 according to the formula. This procedure is to be applied for light curves of the given object measured in all available colours. The final refinement of the period value is done calculating the mean of the values obtained in different colours inversely weighted with their uncertainties.

Using the refined value of period obtained, the parameters of the harmonics are searched for in the following form:

$$m(\phi) = C + A_0 sin(2\pi(\phi + \Phi_0)) + A_1 sin(2\pi(2\phi + \Phi_1)) + ... + A_n sin(2\pi(n\phi + \Phi_n)), \quad (2)$$

where ϕ is the running phase, with respect to the phase at an arbitrary initial epoch, obtained from the moments of individual observations and the fitted period; m is the magnitude at the given ϕ; C is the mean magnitude; A_0, A_1, and A_n are the amplitudes and Φ_0, Φ_1, and Φ_n are the initial phases of the main frequency, the 1st, and the nth harmonics. The values of these parameters are obtained applying a least squares fit to the observed data. The Fourier coefficients are calculated as follows: $R_{ij} = A_i/A_j$ and $\Phi_{ij} = \Phi_i - \Phi_j$. These are meant to be used for the determination of the subclass of the Cepheid and its pulsation mode. The binarity status is determined from the relative amplitudes observed in different colours (Klagyivik & Szabados 2009).

In case of sufficient data points and phase coverage, possible changes in the period and light curve shape are also examined. For the expected number of data points for Gaia time series is usually low, about 70 or even smaller, period changes or even changes in the light curve shape can be pointed out by splitting time series into separate cycles. The data points of the different cycles are compared to each other. To describe period changes the least squares fit to the time series is determined using a period changing linearly with time. This fitting is quite sensitive to the initial values of the parameters, especially the rate of period change. These values are estimated determining the periods of the splitted parts of time series separately. A more complex picture of the period changes can be achieved by generating $O - C$ diagrams of the given light curve data. This method, however, needs longer time series to reach meaningful results.

2.2. Input data for testing
While the algorithms being developed are meant to be used for the processing of future Gaia data, testing of these procedures can only be performed on simulated Gaia data and archive datasets. Appropriate test data must be similar to the expected Gaia data at least in some particular aspects. For this reason the applicability of existing data archives is quite limited. Time series with the same level of photometric uncertainties can only be obtained from simulating Gaia measurements. However, there are archival datasets with time series similar in extent, time sampling, and number of data points. Though the uncertainty of these data are notably higher, this affects the tests only in the sense that errors of the test results will be much higher than expected for Gaia results. Accordingly, while the accuracy of the Gaia results can be estimated only from simulated Gaia data, the detectability of some particular phenomena constrained by mostly the time sampling, length and number of elements of the dataset can be confirmed by testing with archival data. For this purpose Berdnikov (2008) and OGLE III (Soszynski et al. 2008) data are used. The automated procedures were successfully tested on more than 3000 time series in total.

3. Preliminary results
Fourier spectra were generated for all the time series used for testing, and the least squares fit mostly resulted in an accurate period value for the given epoch. In some cases more than one period was found according to the different modes or the different periods of the components of a binary Cepheid. The obtained Fourier spectra showed significant peaks corresponding to the previously published period values. The determination and refinement of the period resulted in an accuracy between 10^{-5}- $10^{-8}\,d^{-1}$ depending on the number of elements and length of the time series, and the accuracy of the observed data. Fourier parameters were also determined. The number of significant harmonics is between 3 and 12 depending on the phase coverage. The

typical value of the residual of light curve fitting was around 0.01 mag. In some cases period change could also be recognised splitting the data into separate cycles, though mostly for time series with larger numbers of elements as it was expected from the Gaia observations. In case of changing period there could not be found a constant period for which data points of all the different cycles overlap each other. For some of these, use of period changing linearly with time resulted in a much better fit to the observed data, even when the length of the time series was not longer than the expected lifetime of Gaia. For some others, mostly time series with length of a decade or longer, no reasonably good fit could be found even using linearly changing period. For such cases generating $O-C$ diagrams showed period changes more complicated than linear. However, significant results could only be obtained for time series with a number of elements much larger than expected from Gaia.

Splitting time series into separate parts can reveal changes in the shape of the light curves, however, because of insufficient data points even the presence of this phenomenon is doubtful. A poor fit with constant period may be caused by the presence of two or more periods in the pulsation. Fourier decomposition is still useful in this case to estimate the rough values of the periods excited simultaneously.

4. Summary

This work concentrates on the development of algorithms for Gaia data processing, the testing of the algorithms developed, and its probable results. Algorithms that can be used to automatically determine and refine periods as well as period changes of Cepheids were described briefly. Further algorithms being developed to characterise possible changes in light curve shape and binarity status of the measured object were mentioned. It was pointed out that the expected characteristics of Gaia data in terms of time sampling, phase coverage, photometric accuracy and the amount of data per object are suitable to provide us important information of many previously known and several thousands of new Cepheid variables with unprecedented accuracy using these algorithms. Periods, Fourier parameters and in many cases period change rates can be determined more accurately than ever. Though, for some Cepheids with slowly changing period Gaia data alone might not be enough to gain useful information, combining with simultaneous ground based observations and archival measurements, these data can also lead to important results.

These studies also help us to learn more about the relations between pulsating stars and stellar evolution, moreover due to the fact that Cepheids are fundamental distance indicators, the cosmic distance scale can also be established more precisely.

5. Acknowledgements

This research is supported by the European Space Agency (ESA) and the Hungarian Space Office via the PECS programme (contract No. 98090).

References

Berdnikov L N 2008 *VizieR Online Data Catalog* **II** 285
Klagyivik P and Szabados L 2009 *A&A* **504** 959
Soszynski I, Poleski R, Udalski A, Szymanski M K, Kubiak M, Pietrzynski G, Wyrzykowski L, Szewczyk O and Ulaczyk K 2008 *Acta Astr.* **58** 163

Journal of Physics: Conference Series **218** (2010) 012027 doi:10.1088/1742-6596/218/1/012027

Linear period dependencies from free parameters in RR Lyrae models

László Molnár and Zoltán Kolláth

Konkoly Observatory, Konkoly Thege út 15-17, H-1121 Budapest, Hungary

E-mail: `lmolnar@konkoly.hu`

Abstract. Detailed analysis of free parameters of the Florida-Budapest turbulent convective stellar model were carried out. A recent explanation of the Blazhko effect ties the observed modulation to convective structure changes inside the star. Before modelling this process, we test relations of growth rates and linear periods to the free parameters. This information is required to model the varying physical characteristics in the code appropriately. Our analysis shows that parameters related to the properties of the turbulent convective zone have the potential to create variations in pulsation periods and amplitudes, however the magnitudes may not be sufficient.

1. Introduction

A new explanation of the still-unresolved Blazhko-phenomenon was recently suggested by Stothers (2006). The model focuses on the cyclic weakening and strengthening of stellar convection caused by turbulent magnetic dynamo, and varying the pulsation properties of the star. Despite the novel approach, no detailed calculations were carried out, whether the mechanism is able to support Blazhko-type amplitude and timescale variations. In his review, Kovács (2009) also points out the lack of elaboration and some inconsistencies of Stothers' idea.

As the proposed mechanism strongly depends on the magnetic field and the turbulent convection inside the star, exact modelling would require a 3 dimensional approach. An overall effect can be introduced to a 1D model as well, by varying the parameters connected to turbulent convection.

The calculations can be further simplified using amplitude equation (AE) formalism (Buchler & Goupil 1984). With the AE method large number of models can be created, but one has to determine the required coefficients. To do this, we started with calculating a few hydrodinamical models to determine growth rates (κ) and amplitudes without introducing any modulation.

2. The model parameters

The Florida-Budapest code involves seven independent dimensionless α parameters connected to turbulent convection. The details of the model can be found in Kolláth & Buchler (2001). Two parameters out of seven, the parameters controlling convective flux (α_c) and eddy viscosity (α_ν) are the dominant ones governing the amplitude of the pulsation. By modulating these two parameters within physically reasonable values the predictions of the Stothers-model can be inspected. As a further control, we determined all seven $\kappa(\alpha)$ and $P(\alpha)$ relations. Yecko et al. (1998) already published tests regarding the dimensionless parameters. Using the same hydrodinamical code to Cepheids, the effects of the convective flux α_c, mixing length α_λ and eddy viscosity α_ν parameters on the shape and location of the linear instability strip and the internal structure of stellar models were investigated. We treat the mixing length differently, with a fixed value ($\alpha_\lambda = 1.5$) and we fine-tune the model with the other parameters because of their internal dependencies.

Properties of our model star are listed in Table 1.

Table 1. Properties of the RR Lyrae model star used in the calculations.

Mass	T_{eff}	Luminosity	Z
0.77 $M_\odot$	6100 K	50 $L_\odot$	10^{-4}

3. Linear periods

The dependence of the linear fundamental period is shown on Figure 1. The most dominant parameter is clearly the parameter governing the convective flux (α_c). The turbulent source function controlled by α_s follows the same trend at low values but it saturates around $\alpha_s \sim 0.4$. Higher values make the model numerically unstable, which clearly shows up in the κ values. Changing the eddy viscosity with α_ν does not affect the stellar structure hence the periods are independent from it. The remaining three parameters have also little if any effect on periods, so for clarity they were omitted from the plot.

Besides the amplitude modulation, Blazhko-variables show phase modulation as well, which can be explained by the variation of pulsation period similar to the amplitude variations. Measuring the magnitude of period changes is not an easy task as it requires good coverage for an entire Blazhko period. A current example is V783 Cyg where the one month long, continuous observations of the Kepler satellite allowed the direct calculation of an instantaneous period that shows period changes with $\sim 0.55\%$ half-amplitude (see Fig. 2 in Kolenberg et al. 2010). The same method applied to four CoRoT RR Lyrae stars revealed variations between $0.1 - 0.7\%$ (Szabó et al. 2009). We estimated period changes from published phase modulation plots and pulsation-modulation frequency values as well. Four stars, SS Cnc (Jurcsik et al. 2006), RR Lyr (Kolenberg et al. 2006), AR Her (Smith et al. 1999) and MW Lyr (Jurcsik et al. 2008) have quite similar values to each other and to V783 Cyg, between $0.4 - 0.6\%$ while for RR Gem (Jurcsik et al. 2006) and DM Cyg (Jurcsik et al. 2009) it is somewhat lower, $\sim 0.15\%$. O-C curves[1] with good coverage over Blazhko cycles (e.g. RR Lyr) confirm these values. Blazhko-periods seem to have little effect on the variation of the pulsational periods: for example the two stars with the longest (RR Lyr, 39 days) and shortest (SS Cnc, 5.3 days) modulations have essentially the same amplitude, $\sim 0.44\%$.

To see how these values compare to our calculations, we indicated variations from our $P = 0.6567\,d$ reference period with $\pm 0.1\%$ (dark grey area) and $\pm 0.5\%$ (light grey area) on Figure 1. The results indicate that large amplitude variations need large variations in the parameters, and the only suitable candidate is the convective flux α_c. We note again that these are linear results where only the eigenvalues of the equilibrium model have been calculated, not the full hydrodinamical code. However, no significant corrections are expected between linear and nonlinear (limit cycle) periods, our first calculations show $0.1 - 0.2\%$ shift only.

[1] Data from the GEOS maxima database: `http://dbrr.ast.obs-mip.fr/`

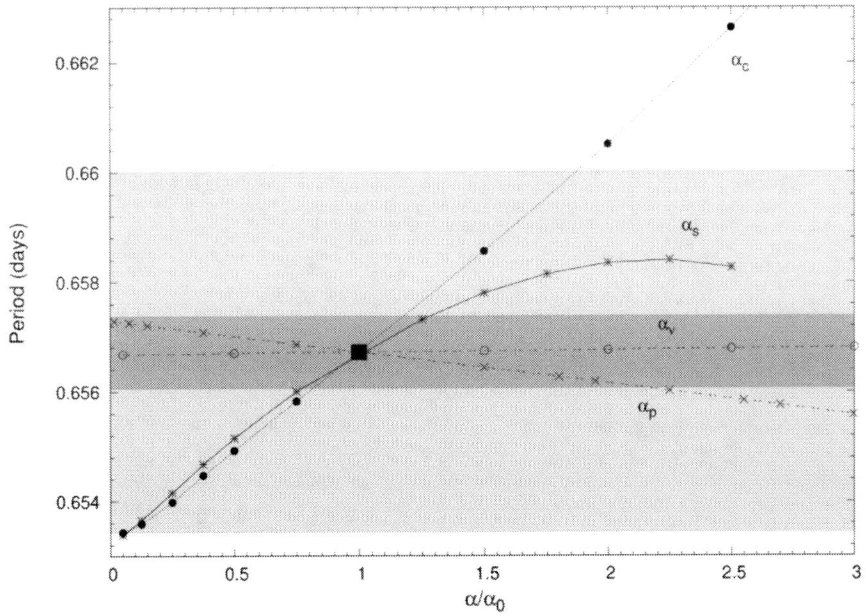

Figure 1. Linear fundamental period values versus normalized α parameters. The black square denotes the reference values of all four parameters. Dark and light grey areas show the $\pm 0.1\%$ and $\pm 0.5\%$ changes to the reference periods respectively.

4. Growth rates and amplitudes

Similarly to the periods, the largest effect on growth rates (κ) come from α_c and α_s, but the eddy viscosity (α_ν) now follows them closely. The other four α parameters also have some, albeit minor effect on growth rates.

We estimated amplitude variations with a simple relation derived from AEs. The simplest form of an AE with a single mode present is $\dot{A} = \kappa A - qA^3$. Considering a limit cycle solution ($\dot{A} = 0$) the amplitude is given by $A = \sqrt{\kappa/q}$. Although the q coefficient is not independent from the other variables, we assume a constant value for our estimate. Variations from the reference value are now simple to compute, $\Delta A/A \sim \sqrt{\kappa}$ as it is shown on Figure 2.

Such very basic estimate suggests a magnitude of $\pm 10 - 20\%$ variations in amplitude-like quantities. Deriving real variations in physical characteristics require running the hydrodinamical code and compute considerably large number of pulsation cycles to sort out all transient features. In the case of α_c, our preliminary results with stellar radius variations show q increasing towards higher α values, making the amplitude-dependence steeper thus allowing slightly higher modulation amplitudes.

5. Conclusions

We investigated RR Lyrae model sequences with the Florida-Budapest code to derive relations between free parameters of the model and linear periods and growth rates of the star. Although our results are only preliminary estimates, these exercises can give us early results regarding the mechanism for the Blazhko-effect proposed by Stothers. The most potent parameters, the convection strength, eddy viscosity and turbulent source function are all connected to the convective structures in the star, in accordance with Stothers' idea. However the actual values seem to reach magnitudes of low-amplitude Blazhko-modulation only. To explain all Blazhko-

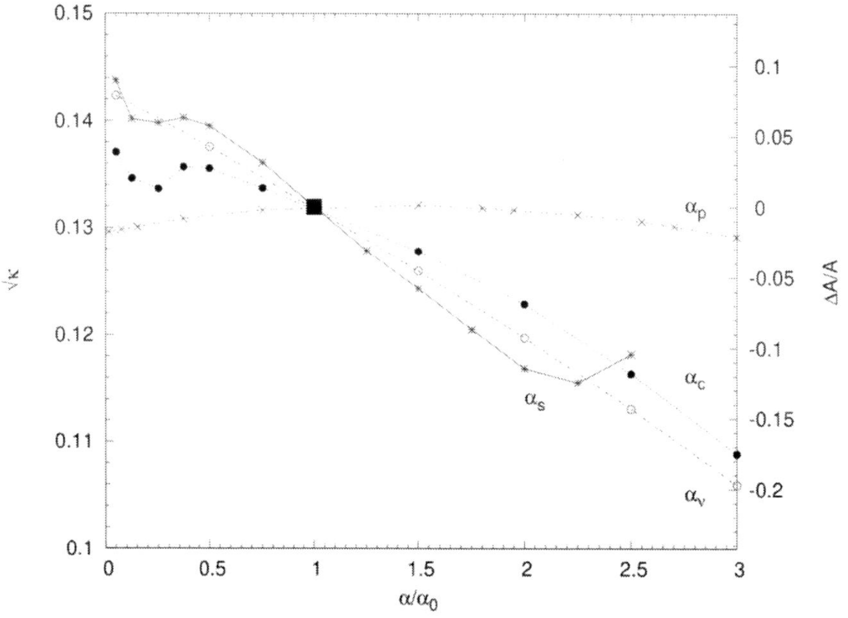

Figure 2. Square roots of growth rates and estimated amplitude-variations versus normalized α parameters.

variables, huge modulation of α parameters has to be introduced, which causes significant changes in the convective structure of the stellar interior. The reality of such large variations over a few weeks in the stellar structure is doubtful.

Our analysis will follow with the more detailed nonlinear model calculations. Modulation properties will be investigated with amplitude equations and a few modulated hydrodinamical model runs to verify the AE results. The parameter space is of course larger than the values we covered here, but these early results already indicate that the mechanism proposed by Stothers faces difficulties.

References

Buchler J R & Goupil M 1984 *ApJ* **279** 394

Jurcsik J, Szeidl B, Sódor Á, Dékány I, Hurta Zs, Posztobányi K, Vida K, Váradi M and Szing A 2006 *AJ* **132** 61

Jurcsik J et al. 2008 *MNRAS* **391** 164

Jurcsik J et al. 2009 *MNRAS* **397** 350

Kolenberg K et al. 2006 *A&A* **459** 577

Kolenberg K et al. 2010 *Preprint* arXiv:1001.0417v1 [astro-ph.SR]

Kolláth Z & Buchler J R 2001 *Stellar Pulsation – Nonlinear Studies* (Astrophysics and Space Science Library Series vol 257) ed M Takeuti and D D Sasselov (Dordrecht: Kluwer Academic Publishers) pp 29-60

Kovács 2009 *AIP Conference Proceedings* **1170** 261

Smith H A, Barnett M, Silbermann N A and Gay P 1999 *AJ 118* 572

Stothers R B 2006 *ApJ* **652** 643

Szabó R, Paparó M, Benkő J M, Chadid M, Kolenberg K and Poretti E 2009 *AIP Conference Proceedings* **1170** 291

Yecko P A, Kolláth Z and Buchler J R 1998 *A&A* **336** 553

5th Workshop of Young Researchers in Astronomy & Astrophysics IOP Publishing
Journal of Physics: Conference Series **218** (2010) 012028 doi:10.1088/1742-6596/218/1/012028

Looking into the heart of a beast: the black hole binary LS 5039

Tamas Szalai[1], **Laszlo L Kiss**[2,3] **and Gordon E Sarty**[4]

[1]Department of Optics and Quantum Electronics, University of Szeged, Dóm tér 9., Szeged H-6720, Hungary
[2]Konkoly Observatory of the Hungarian Academy of Sciences, H-1525 Budapest, P.O. Box 67, Hungary
[3]Sydney Institute for Astronomy, School of Physics A28, University of Sydney, NSW 2006, Australia
[4]Department of Physics and Engineering Physics, University of Saskatchewan, Saskatoon, Saskatchewan S7N 5E2, Canada

E-mail: `szaszi@titan.physx.u-szeged.hu`

Abstract. LS 5039 is a relatively close microquasar consisting of a late O-type star and a compact object (very possibly a black hole) on a highly eccentric orbit with a period of 3.9 days. The high X-ray, gamma-ray and radio luminosity indicate light-matter interaction, which arise from the stellar wind of the primary star accreting toward the black hole. Former examinations suggest that LS 5039 could be a prototype of wind-fed high mass X-ray binaries (WXBs) with diskless main sequence O primaries. Now there is a great chance to better understand the configuration and the physical processes in the exotic system. In July 2009 LS 5039 was followed by the Canadian MOST space telescope to get ultraprecise photometric data in a month-long semi-continuous time series. Parallel to this, we have taken simultaneous high-resolution optical spectra using the 2.3m ANU telescope of the Siding Spring Observatory, supplemented with further data obtained in early August 2009 with the same instrument. Here we present the first results from the new echelle spectra, which represent the best optical spectroscopy ever obtained for this intriguing system. We determined fundamental orbital and physical parameters of LS 5039 and examined the configuration and the circumstellar environment of the system via radial velocity measurements and detailed line-profile analysis of H-Balmer, He I and He II lines.

1. Introduction

LS 5039 (V479 Sct), classified as a high-mass X-ray binary (HMXB) by Motch et al. (1997), is one of the most studied members of its class. Based on VLA data, Martí et al. (1998) discovered a persistent non-thermal radio counterpart, associated with mildly relativistic radio jets (Paredes et al. 2000, 2002). In addition, a very high energy (VHE) γ-ray source was also found by the CGRO/EGRET (Paredes et al. 2000) and HESS (Aharonian et al. 2005) surveys. The observed features make LS 5039 a prominent member of the so-called gamma-ray binaries.

The optical companion of the system is a bright (V = 11.2), O6.5V((f)) type star classified by Clark et al. (2001) and McSwain et al. (2001, 2004). McSwain et al. (hereafter M04) presented the first orbital parameters of the binary system. Casares et al. (2005; hereafter C05) performed a detailed spectroscopic analysis of LS 5039, and calculated a new set of orbital and physical parameters. They found an orbital period of 3.906 days, which has been confirmed by other studies: for example, variability also in X-rays (Bosch-Ramon et al. 2005, Takahashi et al. 2009), at GeV (Abdo et al. 2009) and TeV energies (Aharonian et al. 2006) occurs with the same period, suggesting orbital modulation of the high-energy emissions. Very recently, Aragona et al. (2009; hereafter A09) – using the old and some new RV points – also confirm the value of the orbital period and refined the parameters.

We have organised a simultaneous campaign of MOST space photometry and ground-based optical spectroscopy, which was executed in July 2009. The main goal of the project was an investigation of the system parameters (based on a new, independent, high-resolution data set), as well as an attempt to detect clumpiness in the stellar wind of the O-type star. Here we report on the first results of our spectroscopic analysis, while the detailed discussion of the results of the coordinated campaign will be presented elsewhere (Sarty et al., in prep).

2. Observations and data reduction

Spectroscopic observations were carried out on four nights between 2009 July 8-11 (parallel to MOST observations) and on three nights between August 1-3, using the 2.3-m telescope of the Australian National University (ANU) equipped with an echelle spectrograph. In total, the obtained 118 spectra cover almost 40 hours and ensure a good sampling of the whole orbit. The integration times were between 900-1200 s, while the spectra covered the whole visual range between 3900 Å and 6720 Å . The nominal spectral resolution is $\lambda/\Delta\lambda \approx 23\,000$ at the Hα line, with typical SNR of about 100 in 1 hour integration.

All data were reduced with standard IRAF[1] tasks, including bias and flat-field corrections, cosmic ray removal, extraction of the 27 individual orders of the echelle spectra, wavelength calibration, and continuum normalization.

3. Analysis and results

3.1. Radial velocities

To measure radial velocities we first generated one hour long average spectra to get higher S/N – one hour corresponds to 0.01 orbital phase, hence negligible phase smearing appears in the phased radial velocity data. The data were phased with an orbital period of 3.906 days (C05) and our RV curve confirms that this is indeed the correct value.

To generate the final RV diagrams we used average velocities of H I (Hα, Hβ, Hγ, Hδ, λ3835), He I (λ4471, λ5875) and He II (λ4200, λ4686, λ5411) lines; there are several other H and He lines in the wavelength region of our spectra, but they were too noisy or blended to use them for velocity determination. The typical measurement error is $\pm$10-15 km s^{-1}, which is partly caused by the observational noise, partly by the large rotational velocity of the O-type star (v sin i = 113 $\pm$ 8 km s^{-1}, C05).

Our results indicate that there is a large systematic shift between He II, He I and H Balmer lines: the H I and He I lines are always blueshifted by about 20 km s^{-1} in respect to the He II lines, which is a characteristic signature of a strong stellar wind. The photosphere of the O-type star is too hot for the H I and He I lines ($\approx$ 40 000 K), hence these are generated in the cooler outer regions. Interestingly, this velocity difference between the different species and ionisation states has not always been taken into account: C05 also examined this effect (but they found smaller differences between velocities), but it was simply neglected by A09.

We also measured equivalent widths of several interstellar lines to estimate the rate of interstellar reddening (see Section 3.2). During that we found redshifted satellite absorptions in the Ca II K and Na I D1& D2 lines with a radial velocity around +60 km/s (more precisely,

[1] IRAF is distributed by the National Optical Astronomy Observatories, which are operated by the Association of Universities for Research in Astronomy, Inc., under cooperative agreement with the National Science Foundation.

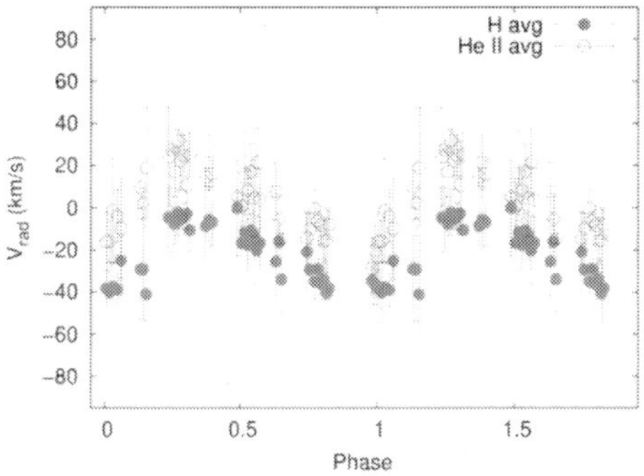

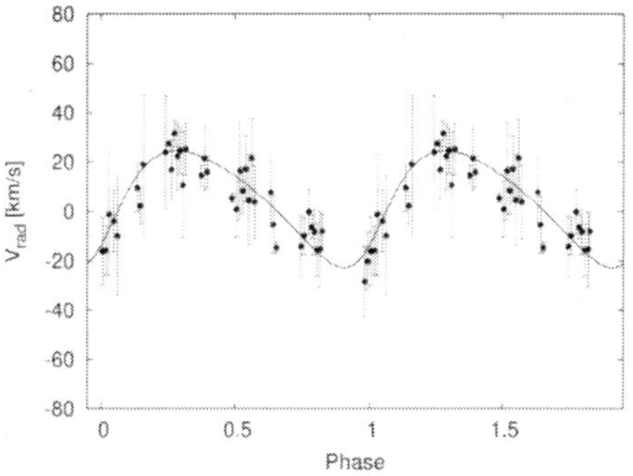

Figure 1. *Left*: Radial velocities based on H Balmer and He II lines. *Right*: The best-fitting curve to radial velocities of He II lines.

+58.4 ± 2.2 km s^{-1} by Ca K, +62.9 ± 2.3 km s^{-1} by Na D1 and +61.9 ± 1.7 km s^{-1} by Na D2). These satellite lines may belong to a formerly unknown galactic Intermediate Velocity Cloud (IVC; see a review in Wakker & van Woerden 1997).

3.2. Orbital and physical parameters

In the following analysis we adopted $T_{\rm eff}$ = 39 000 ± 1000 K and log g = 3.85 ± 0.10 for the stellar companion (C05). To determine interstellar reddening we used Na I D1 (EW = 0.70 ± 0.02 Å ; see for ref. Munari & Zwitter 1997), DIB λ5780 and DIB λ6613 (EW = 0.55 ± 0.05 Å and 0.18 ± 0.02 Å , respectively, see for ref. Cox et al. 2005, Fig. 4.) lines, and we have got $E(B-V)$ = 1.2 ± 0.1, which is in very good agreement with previous results (Ribó et al. 2002, M04). Using other photometric parameters (R, M_v, see C05 and references therein) we also confirmed the distance (d = 2.5 ± 0.1 kpc), stellar radius ($R_0 = 9.3^{+0.7}_{-0.6}$ R$_\odot$) and mass of the O star ($M_0 = 22.9^{+3.4}_{-2.9}$ M$_\odot$) published by C05.

Radial velocity curves were modelled using the 2003 version[2] of the Wilson-Devinney (WD) code (Wilson & Devinney 1971, Wilson & van Hamme 2003). We note that during the analysis we only used our own velocity points, which are independent from any previous data in the literature, and constitute the highest resolution, homogenous spectral dataset ever used to get the orbital solution of LS 5039.

The determined orbital parameters are shown in Table 1, compared to results of C05 and A09. The ones published by C05 were based only on velocity points related to He II lines, while A09 combined the velocities of every available H, He I, and He II lines (see our cautionary notes above).

In general, our parameters are close to previous solutions, but there are some differences. The value of V_γ is definetly higher in C05 than in A09 or in our paper. The possibility of a real change in this parameter during only some years is very small, so the reason of the difference could be effect from different data treatment in of C05. The other important difference is our smaller value of e, what we explain with the smaller scatter and better coverage of our dataset.

[2] ftp://ftp.astro.ufl.edu/pub/wilson

Table 1. Orbital parameters of LS 5039.

Parameter	C05 (He II)	A09	This paper (He II)
T_0 (HJD−2450000)	1943.09 ± 0.10	2825.99 ± 0.05	5017.08 ± 0.06
P_{orb} (d)	3.90603	3.90608	3.906 (adopted)
e	0.35 ± 0.04	0.34 ± 0.04	0.24 ± 0.08
ω [°]	225.8 ± 3.3	236.0 ± 5.8	237.3 ± 21.8
V_γ [km s^{-1}]	17.2 ± 0.7	4.0 ± 0.3	3.9 ± 1.3
K_1 [km s^{-1}]	25.2 ± 1.4	19.7 ± 0.9	23.6 ± 4.0
a_1 sin i [R$_\odot$]	1.82 ± 0.10	1.44 ± 0.07	1.77 ± 0.15
$f(m)$ [M$_\odot$]	0.0053 ± 0.0009	0.0026 ± 0.0004	0.0049 ± 0.0006
rms of fit [km s^{-1}]	9.1	7.1	6.2

3.3. *Circumstellar matter*

To infer mass loss rate of the O star and the properties of the circumstellar matter, we determined the equivalent widths of each H and He line and also their variability during the orbit, which could be a good indicator of the physical processes taking place in the stellar wind. We executed the measurements on the average spectra summarized from one hour long exposure times, but we plotted the daily average values of *EW*s on final diagrams (to better see the trends).

For the Hα line we found that EW changes between 2.50 and 2.85 Å , and its variability shows a weak correlation with the orbital period. The main averege value (2.70 ± 0.12 Å) agree within uncertainties with the result of C05 (2.8 ± 0.1 Å), except that they found that value is stable during the orbit of the binary (which reason could be the relatively low resolution of their spectra). Our result is also consistent with the EW values measured in last ten years (Bosch-Ramon et al. 2007). Using the results of Puls et al. (1996, and references therein) we could estimate the mass loss rate from EW of Hα line. To do these calculations we also had to use some other parameters: $R_0 = 9.3^{+0.7}_{-0.6}$ R$_\odot$ and $T_{\rm eff}$ = 39 000 ± 1000 K of the O star (C05), terminal velocity (V_∞ = 2440 ± 190 km s^{-1}, M04) and wind velocity law exponent (β = 0.8, M04). We found that the mass-loss rate of the stellar companion is around 3.7 × 10^{-7} M$_\odot$ yr^{-1} by largest absorption (which correspond with the lower limit), while we could give 4.8 × 10^{-7} M$_\odot$ yr^{-1} for upper limit. These values also confirm the results of C05.

There are two lines (Hβ and He I λ5875) by which the values of EW are more strongly modulated by orbital period. The lowest absorption is around φ ∼0.75 by H lines, and φ ∼0.65 by the mentioned He line (Fig. 2, left side), close to the expected phase φ ∼0.7 when the compact object is between us and the stellar companion (inferior conjunction).

By H Balmer lines we also tried to determine the phase dependence of *EW*s only of emission components. Because of the relatively small S/N it was not performable to follow hour-to-hour changes in line profiles, so therefore we generated difference spectra subtracted Doppler-corrected daily average spectra from their main average to calculate *EW*s of remaining emission components. Using these values quantitatively would be an excessive implementation of results, but the relative, phase dependent changes are examinable.

The three diagrams (Fig. 2, right side) are very similar to each other, with a significantly exceeding point (which means maximal emission here) near inferior conjunction (φ ∼0.65). It agrees with the diagrams shown on the left side of Fig. 2, where lowest absorption is also at that phase. We can rule out any technical artifacts which could produce the similarity of the three diagrams, because the lines are in different echelle orders, so they went independently through the different steps of data reductions.

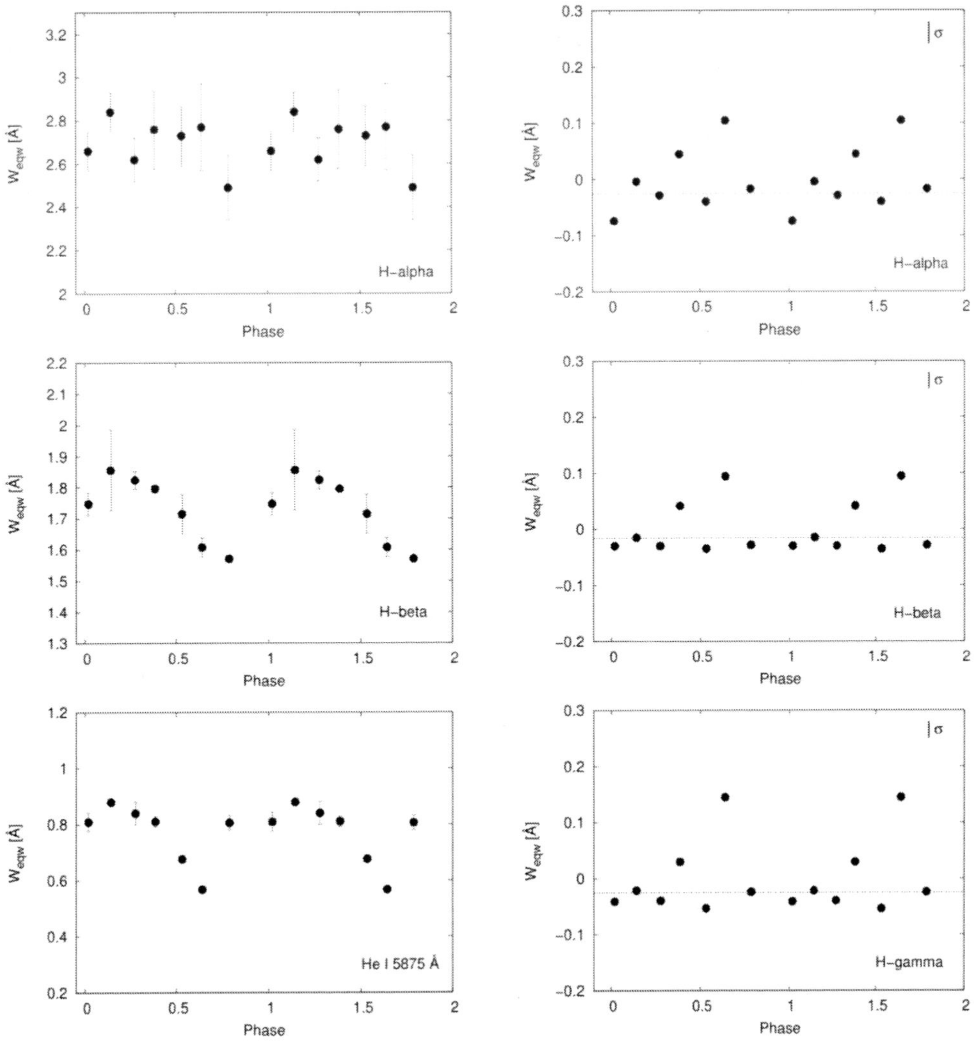

Figure 2. *Left*: Variability of equivalent widths of Hα, Hβ and He I λ5875 lines during the orbit. *Right*: Variability of equivalent widths of the emission components of Balmer lines during the orbit of LS 5039. In these diagrams larger numbers correspond to larger emission rates. There is a point by each line with a significantly high value (3, 3 and 5 σ above average, respectively) at $\varphi \sim$0.65.

4. Summary

In our work we presented the first results of a coordinated campaign consisting of parallel space photometry and high-resolution ground-based spectroscopy of the the X-ray (gamma-ray) binary LS 5039. Detailed analysis of spectroscopic data (which incarnate the highest resolution, homogenous dataset ever obtained from that object) allowed us to determine the orbital parameters of the system being mainly consistent with former solutions, but we also found some differences (emphasizing the significant velocity shift between H I, He I and He II lines, and the smaller value of eccentricity).

To reveal the properties of the circumstellar environment, we first measured the equivalent widhts of available H and He lines during the orbit. We found that Hα, Hβ and He I λ5875 lines are unambiguously modulated by orbital period – the clear emission components of H Balmer lines show similar phase dependence, which seems to confirm that it is a real effect. From the average EW of Hα line we estimated the mass loss rate of the O star and got similar value to former results.

The complete analysis of our observations about LS 5039 will be presented in a latter paper (Sarty et al., in prep), including discussion about the circumstellar environment and possible mass ranges of the compact object.

Acknowledgments

This work has been supported by the Australian Research Council, the University of Sydney, the Hungarian OTKA Grant K76816 and the "Lendület" Young Researchers' Program of the Hungarian Academy of Sciences.

References

Abdo A A et al 2009 *ApJ* **706** L56
Aharonian F A et al 2005 *Science* **309** 746
Aharonian F A et al 2006 *A&A* **460** 743
Aragona C, McSwain M V, Grundstrom E D, Marsh A N, Roettenbacher R M, Hessler K M, Boyajian T S and Ray P S 2009 *ApJ* **698** 514 (A09)
Bosch-Ramon V, Paredes J M, Ribó M, Miller J M, Reig P and Martí J 2005 *ApJ* **628** 388
Bosch-Ramon V, Motch C, Ribó M, Lopes de Oliveira R, Janot-Pacheco E, Negueruela I, Paredes J M and Martiocchia A 2007 *A&A* **473** 545
Casares J, Ribó M, Ribas I, Paredes J M, Martí J and Herrero A 2005 *MNRAS* **364** 899 (C05)
Clark J S, Reig P, Goodwin S P, Larionov V M, Blay P, Coe M J, Fabregat J, Negueruela I, Papadakis I and Steele I A 2001 *A&A* **376** 476
Cox N L J, Kaper L, Foing B H and Ehrenfreund, P 2005 *A&A* **438** 187
Martí J, Paredes J M and Ribó M 1998 *A&A* **338** L71
McSwain M V, Gies D R, Huang W, Wiita P J and Wingert D W 2004 *ApJ* **600** 927 (M04)
McSwain M V, Gies D R, Riddle R L, Wang Z and Wingert D W 2001 *ApJ* **558** L43
Motch C, Haberl F, Dennerl K, Pakull M and Janot-Pacheco E 1997 *A&A* **323** 835
Munari U and Zwitter T 1997 *A&A* **318** 269
Paredes J M, Martí J, Ribó M and Massi M 2000 *Science* **288** 2340
Paredes J M, Ribó M, Ros E, Martí J and Massi M 2002 *A&A* **393** L99
Puls J et al 1996 *A&A* **305** 171
Ribó M, Paredes J M, Romero G E, Benaglia P, Martí J, Fors O and García-Sánchez J 2002 *A&A* **384** 954
Takahashi T, Kishishita T, Uchiyama Y, Tanaka T, Yamaoka K, Khangulyan D, Aharonian F A, Bosch-Ramon V and Hinton J A 2009 *ApJ* **697** 592
Wakker B P and van Woerden H 1997 *ARA&A* **35** 217
Wilson R E and Devinney E J 1971 *ApJ* **166** 605
Wilson R E and van Hamme W 2003 *Computing Binary Stars Observables* ver 4 (Gainesville: University of Florida)

Journal of Physics: Conference Series **218** (2010) 012029 doi:10.1088/1742-6596/218/1/012029

Testing the global flow reconstruction method on coupled chaotic oscillators

Emese Plachy[1] and Zoltán Kolláth[2]

[1]Department of Astronomy, Eötvös University, Pázmány Péter sétány 1/a, H-1117 Budapest, Hungary

[2]Konkoly Observatory, Konkoly Thege Miklós út 15-17, H-1121 Budapest, Hungary

E-mail: E.Plachy@astro.elte.hu

Abstract.
Irregular behaviour of pulsating variable stars may occur due to low dimensional chaos. To determine the quantitative properties of the dynamics in such systems, we apply a suitable time series analysis, the global flow reconstruction method. The robustness of the reconstruction can be tested through the resultant quantities, like Lyapunov dimension and Fourier frequencies. The latter is specially important as it is directly derivable from the observed light curves. We have performed tests using coupled Rössler oscillators to investigate the possible connection between those quantities. In this paper we present our test results.

1. Introduction

If the timescale of the modulation in semiregular data is comparable to the period, that may indicate chaotic dynamics in the background. Radially pulsating semiregular variable stars often show such modulation in their light curves due to nonlinear nonadiabatic properties of their oscillation. Using the global flow reconstruction method, evidence has been found for low dimensional chaos in several cases. These investigations suggest that irregular pulsation occurs as the result of energy exchange between coupled modes (Buchler et al. 1996, Kolláth et al. 1998, Buchler et al. 2004). To model the interaction of the modes we have generated artificial data of coupled Rössler oscillators. It is well known that Rössler oscillators have chaotic solution for certain parameter combinations. This applies for two coupled Rössler oscillators as well. So we have based our tests on a priori known chaotic time series with two interacting oscillation modes. As the chaotic semiregular stars have two groups of peaks in their Fourier spectra, our examinations concern the one dimensional output of the coupled Rössler oscillators that clearly show the presence of the two frequency groups.

2. Coupled Rössler Oscillators

We have generated our artificial data with the following set of differential equations, where the first three equations describe one oscillator and the last three equations describe the other one.

$$\dot{x}_1 = -a(x_2 + x_3), \quad \dot{x}_2 = a(x_1 + bx_2) + ex_5, \quad \dot{x}_3 = a(c + x_3(x_1 - d))$$
$$\dot{x}_4 = -f(x_5 + x_6), \quad \dot{x}_5 = f(x_4 + gx_5) + jx_2, \quad \dot{x}_6 = f(h + x_6(x_4 - i))$$

Ten parameters are involved, a and f refer to the oscillation frequencies, while e and j assure the coupling through x_5 and x_2 variables. By varying the parameters we have constructed different data sets. In our tests we have used the 1D time series of x_1 variable. Gaussian noise was also added to the test signals to generate data similar to real observations.

3. The Global Flow Reconstruction Method

Our nonlinear analysis starts with the construction of a strobed representation of the trajectory of the system in the reconstruction space. Using the sequence with equal time spacing $s(t_n)$ we produce the so called 'delay vectors' $\mathbf{X}(t_n) = s(t_n), s(t_n - \Delta), s(t_n - 2\Delta), ..., s(t_n - (d_e - 1)\Delta)$, where Δ is the 'time delay' and d_e is the embedding dimension of the reconstruction space. We assume that there exists a map $\mathbf{F}$ that evolves the trajectory in time by connecting the neighboring points, $\mathbf{X}^{n+1} = \mathbf{F}(\mathbf{X}^n)$. We fit the map $\mathbf{F}$ in polynomial form. By iterating the constructed $\mathbf{F}$ map we can produce synthetic signals that can be compared to the original data. For comparison different representations are useful, the Fourier spectrum, Broomhead-King (BK) projection (Broomhead & King, 1987), and the data set itself. Synthetic and original data must be in accordance in all representations to be acceptable as a good robust map. By varying the 'time delay' Δ, the embedding dimension d_e, the smoothing parameter σ and the added noise intensity ξ we can identify a good robust map region in the parameter space. Good reconstructions provide approximated quantitative information of the system, like Lyapunov exponents and dimension. For detailed description of the method we refer to Serre et al. (1996).

4. Discussion

We have applied the global flow reconstruction method on the test data. We have used 5 and 6 dimensional reconstruction space ($d_e = 5, 6$). The order of the polynomial fit has been fixed at 4 ($p = 4$). Smoothing parameter ($\sigma = 0.000, 0.001, 0.002$), noise intensity ($\xi = 0.000, 0.001, 0.002$) and 'time delay' ($\Delta = 4 - 25$) characterize our examined parameter regime. The iterated synthetic data of the successful reconstruction maps have been compared to the original test data.

Considering we deal with chaotic data, we can not expect exact accordance comparison method, only strong similarity. We present a subsample of tests on Figure 1. We compared the test dataset, a different realisation of the test dataset and two reconstructed synthetic signals. The difference of *Test D* lies in the initial value of the iteration. *Syn 1* is the synthetic signal of the best reconstruction and *Syn 2* is a typical synthetic time series of the examined parameter space. Comparison of a fraction of datasets and the BK projections has revealed that reconstructed signals do not differ from the test data more than a different realisation of the test data itself. Figure 2 displays the Fourier spectra of the same datasets, that show two groups of frequencies, a larger group around 0.0035, and a smaller group at 0.0128. The frequency spectra of synthetic signals are in good correspondence to the test data as well, however the signals of typical reconstructions show different periodicities causing noticeable shifts of frequency peaks.

Our further investigations have focused on the frequency peaks. As chaotic oscillation shows temporal modulation, the oscillation frequencies do not appear as single strong peaks in the Fourier spectrum, but as a forest of peaks in a certain frequency range. Different realisations of the same system display different behaviour in their frequency spectra, like different location of the highest amplitude peaks. Therefore spectra can be compared only in statistical sense.

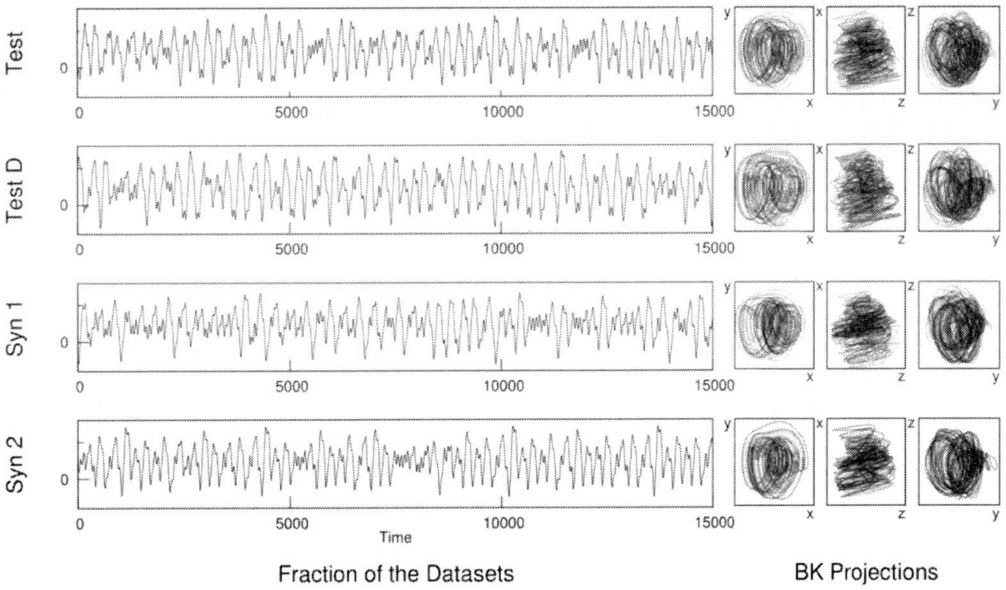

Figure 1. Comparison of Rössler test data and synthetic solution by different representations.

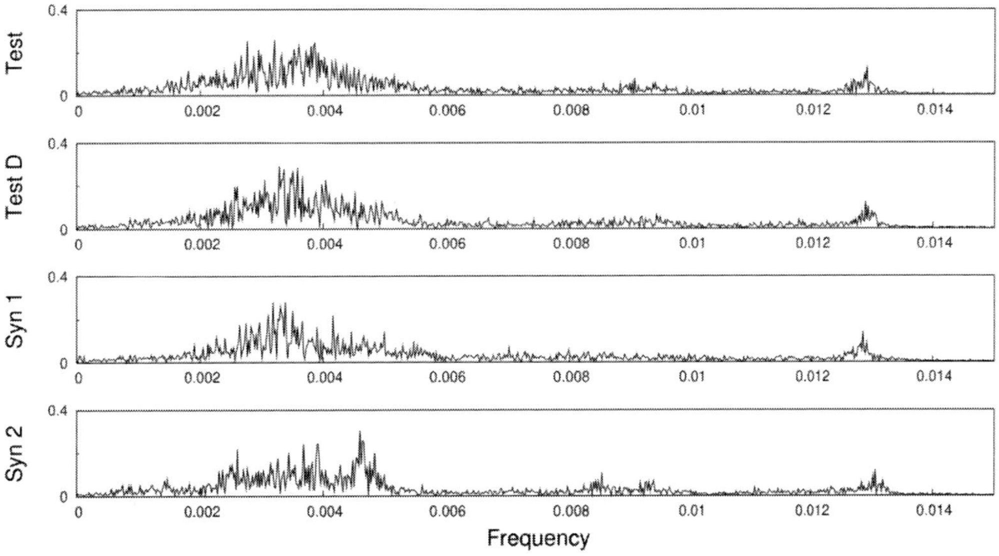

Figure 2. Comparison of Rössler test data and synthetic reconstructions by Fourier spectra.

We have applied Gauss curve fitting on the frequency spectra to perform a reliable comparison by determining the centre position of the Gaussian envelope of the test dataset and the synthetic signals. With this manner we have obtained 0.003469 ± 0.000013 and 0.012875 ± 0.000008 frequency values for the original coupled Rössler oscillator. The frequency values of synthetic signals are 0.00361 ± 0.00022 and 0.01289 ± 0.00016 in 5D, and 0.00359 ± 0.00015 and 0.01288 ± 0.00006 in 6D reconstruction. The central frequency of Gaussian fit to the Fourier spectra is a well defined and useful quantity. Our analysis revealed that the original and synthetic values of the central frequencies differ significantly.

We have tested the connection between the frequencies and the Lyapunov dimensions to investigate the correlation between the errors of the quantities (Figure 3). The Lyapunov dimension value of the test Rössler data has been calculated to be 3.53 ± 0.07, while synthetic signals have 3.03 ± 0.30 (5D) and 3.11 ± 0.31 (6D) values. Figure 3 clearly shows the large scatter in Lyapunov dimensions and in both frequency groups. The majority of the Lyapunov dimension values are below the original ones.

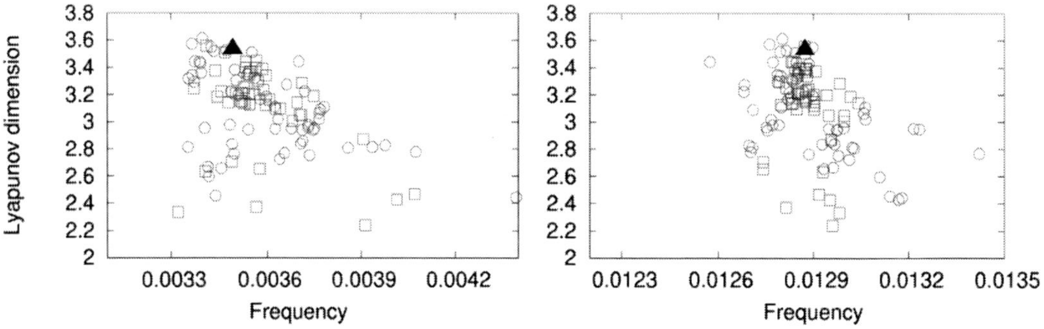

Figure 3. Lyapunov dimension versus centre position frequency of the Gaussian envelope. 5D: circle, 6D: square, test dataset: triangle.

5. Conclusion

We have succeeded in reconstructing artificial datasets with Lyapunov dimension higher than 3, that was composed by coupling two oscillations. Our tests suggest however, that the global flow reconstruction method do not reconstruct the exact frequencies and the Lyapunov dimension of the original data. The errors of these quantities do not correlate with each other. The lack of correlation between the examined measures confirms that both quantities must be used to determine the robustness of the reconstruction, or to strain off the false reconstructed signals. We can not use the average values of the measures either as they show systematic deviation beside their large scatter. This results should be taken into consideration when real world observations are analysed and interpreted. Possible direction of further investigations can be the development and test of a reconstruction method which guarantees the results with smaller frequency discrepancy.

References

Buchler J R, Kolláth Z and Serre T 1996 *ApJ* **462** 489
Buchler J R, Kolláth Z and Cadmus R R Jr 2004 *ApJ* **613** 532
Broomhead D S and King G P 1987 *Physica D* **20** 217
Kolláth Z, Buchler J R, Serre T and Mattei J 1998 *A&A* **329** 147
Serre T, Kolláth Z and Buchler J R 1996 *A&A* **311** 833S

AUTHOR INDEX

Institute of Physics
Dirac House, Temple Back
Bristol BS1 6BE UK

CURRAN ASSOCIATES INC
proceedings.com